Fish
and Shellfish
Farming in
Coastal Waters

P H Milne

Fish
and Shellfish
Farming in
Coastal Waters

P H Milne

Fishing News Books Ltd
Farnham. Surrey, England.

First published 1972
Reprinted 1979

ISBN 0 85238 0224

Printed by The Whitefriars Press Ltd., London and Tonbridge.

Preface to 1979 Edition

Since the original publication of this book in 1972 there has been a considerable upsurge of interest in aquaculture throughout the world with a significant increase in production and development.

This book deals mainly with the engineering aspects for site selection, design and construction, and these guidelines have not altered radically over the years. Facilities do change with time and the research enclosures constructed by the author (Fig. 103) have now been dismantled.

However, on the biological side there have been many new developments, discoveries and transplantations between continents. Spectacular advances have been made especially in salmon farming and ranching in Canada, North America, Norway and Scotland. Eel farming with its main roots in Japan has now spread to the Philippines and Europe. In the United Kingdom there are three new farms now in operation: at Hinkley Point Power Station in England and at both Tomatin Distillery and J. & P. Coats Mills (Paisley) in Scotland, where in each case heated waste water is used for the ongrowing of the eels.

On the marine farming front, advances have been made at Hunterston (Fig. 47) in the farming techniques for sole and turbot with a ready commercial market for plate-size sole.

In the future, aquaculture will play an important part in the production of protein for the popula-tions of Latin America and the South China Sea. To further the development of aquaculture, both fresh and brackish water in the latter area, SEAFDEC (South East Asia Fisheries Development Centre) was established with its headquarters at Manila in the Philippines, which I visited in 1973. The other countries involved in SEAFDEC are Japan, Malaysia, Thailand, Singapore and Vietnam.

Turning to more exotic species, advances have also been made in lobster farming in Maine (U.S.A.) and crayfish farming in Sweden and Louisiana (U.S.A.).

Readers wishing to keep up to date with both the engineering and biological developments in aquaculture, like those mentioned above, will find regular contributions in the quarterly journal, "Fish Farming International", published by Arthur J. Heighway Publications Ltd, London.

I have been greatly encouraged by the interest shown by readers' letters since this book first appeared in 1972 and I hope it will continue to give inspiration to many more budding fish farmers in the future.

DR P. H. MILNE
January 1979

Preface to First Edition

I became interested in fish and shellfish farming in coastal waters in 1964 with my involvement in the design, construction and development of sea farming enclosures in Scotland. In the Department of Civil Engineering at the University of Strathclyde in Glasgow, there has been noted a steady increase since 1965 in the number of enquiries about methods for the design and construction of coastal enclosures for sea farming. Last year (1971) I received over one hundred requests for information and assistance, originating not only from other research workers, but also from individuals and large commercial concerns interested in the potential development of farming in coastal 'waters.

Consequently it gave me much pleasure to receive the invitation from the publishers to prepare this book for prospective marine farmers to advise and collate for them the latest developments in design and construction techniques used in sea farming throughout the world.

In this book, reference is made, wherever possible, to published scientific literature so that if further details are required the source can be traced. To keep the book reasonably concise I have principally mentioned only the sea farming developments with immediate commercial prospects. For this reason I hope those whose work or area has not been mentioned will understand.

It has been my good fortune to receive very generous assistance from fellow research workers and from fish and shellfish farmers throughout the world, too numerous to mention individually, who provided reports, drawings and photographs to illustrate the text. Much of the information has been obtained by personal contacts on visits to fish and shellfish farms, and by correspondence, and I am grateful to them for permission to refer to their hitherto unpublished results.

I am greatly indebted to the late Professor W. Frazer and to my former colleague, Dr J. H. Allen, for inviting me in 1964 to join the Oceanography Section of the Department of Civil Engineering at the University of Strathclyde to carry out research into the design and construction of facilities for sea farming. In my present post of Lecturer, at the University of Strathclyde, I am grateful to Professor D. I. H. Barr for his continued help and encouragement in publishing the results of my research work.

The British White Fish Authority and the Natural Environment Research Council are also to be thanked for sponsoring my work from 1964 till 1971. I would like to thank all my colleagues during this period in: the Aberdeen Laboratory of the Department of Agriculture and Fisheries for Scotland; the Dunstaffnage (former Millport) Laboratory of the Scottish Marine Biological Association; and the Fish Cultivation Units of the White Fish Authority at Ardtoe, Faery Isles and at Hunterston for their assistance and co-operation during my experiments.

Grateful acknowledgement is made to the respective Editors of Hydrospace, Marine Research, Oceanology International '72 and World Fishing for granting permission to reproduce sections of my papers, figures and photographs previously published in their journals.

Finally, this book could not have been prepared without the wholehearted encouragement and support of my wife, Helen, whom I also thank most sincerely for typing the complete manuscript. I also wish to thank my father-in-law, Mr R. A. Hunter, for the benefit of his advice and experience, from his own previous research publications, in the preparation of this book, and for reading over the entire manuscript prior to publication.

Department of Civil Engineering,
University of Strathclyde,
John Anderson Building,
107 Rottenrow,
Glasgow, G4 0NG,
Scotland.

P. H. MILNE
January 1972

Contents

5 THE FUTURE

List of Figures

1
GENERAL

Chapter 1
Introduction

Fish and shellfish farming has a very long history. For example the Japanese farmed oysters on intertidal stretches of the shore around 2000 B.C. Aristotle mentions the cultivation of oysters in Greece and Pliny gives details of Roman oyster farming from 100 B.C. Again, fresh and brackish water fish farming over the centuries is fully authenticated, the first treatise on fish culture being written in China by Fan Li in 475 B.C. (Dill, 1967).

In comparatively recent times, due to the industrial revolution and the rise in world population, there has been an increase in the discharge of waste products and industrial effluents into the sea. Unfortunately this now apparent and recognised pollution has wiped out large areas of natural oyster beds and destroyed some coastal fisheries.

To combat this depletion of stocks, new methods of fish and shellfish cultivation have been developed, not only with the traditionally farmed species of shellfish, but also with more highly valued species of fish which command a wider market. The most promising locations for these developments are in the sheltered bays and estuaries of our coastal waters, Fig. 1. In the selection of a site for fish and shellfish farming it is essential that thorough preliminary surveys are conducted to determine the stability of the existing marine environment.

Much has been said in the past about how food from the sea could alleviate the critical protein shortage resulting from man's failure to keep pace with the nutritional needs of the exploding population (Lucas, 1966). Yet the traditional harvest of the sea is becoming increasingly difficult to gather and fishing has to be carried out at greater expense further afield. In addition there is the problem of over-fishing which is now even alarming the French fishing fleets (Anon., 1971h). It is interesting to note that the latest Food and Agriculture Organisation of the United Nations (FAO) figures for the world fish catch for 1969 show a fall of 2% from 1968. The totals for 1968 and 1969 are $64 \cdot 3 \times 10^6$ and $63 \cdot 1 \times 10^6$ tonnes respectively (FAO, 1971).

It is believed that a possible solution to this difficult and complex problem may be found in the advance of scientific and practical methods of fish and shellfish culture for farming in coastal waters (Yonge, 1966a; Cole, 1968, 1972). Hence the most beneficial and promising results for the production of food by the fisheries of the western world may lie in the establishment of large-scale marine farming operations in the sea (Hester, 1966). Consequently, the prospective marine farmer has a great opportunity to establish a successful farm (Iversen, 1968).

Despite the long standing development of fresh and brackish water farming (Schuster, 1949; Hickling, 1962, 1968, 1970), it was not until the end of the nineteenth century when there was a general upsurge in oceanography, that the cultivation of marine organisms was intensified. Even with increasing interest in the cultivation of marine species in the twentieth century, less than 2% of all known marine organisms can at present be reared throughout their whole life cycle under controlled conditions (Kinne, 1970). This is not to say that it is impossible to cultivate the other 98%, but that past methods which have often yielded negative results have been due to incomplete methods, lack of equipment, and financial support. Hence the ultimate cultivation of marine organisms involves basic and applied research in all its ecological, engineering and economical aspects.

The most commonly cultivated marine organisms are the molluscs, such as oysters and mussels, which are filter feeders and occupy low positions in the marine food pyramid (Steele, 1970). The carnivorous crustaceans and fishes which require secondary or tertiary food web links are at the top of the marine food pyramid and cannot be farmed as efficiently in terms of food conversion. However, it is this latter group that is becoming of increasing economic importance, and many government laboratories throughout the world have research programmes to study their economic cultivation.

Fig. 1 Typical sheltered inlet of our coastal waters. The inner area protected from the offshore breakers at Ardtoe on the west coast of Scotland was chosen in 1965 by the British White Fish Authority for an intertidal pond. (*Credit—P. H. Milne*)

The numerous varieties of marine organisms at present under cultivation in different parts of the world are considered in Chapter 2. Not only are those species that are being commercially farmed included, but mention is also made of those potentially marketable species which at present are undergoing pilot experimentation. Their inclusion in this book is to aid and advise the prospective marine farmer where current research is in progress since today's research is often tomorrow's commercial technique.

Depending on the topography of the coastline, the climate, natural food sources, water exchange, and the species to be cultivated, sea farms can be operated in various ways. Sedentary organisms can be farmed in the intertidal and seabed zones without the need for fences or enclosures, but with crustacea and fish this is obviously essential. Various coastal zones can be selected and successfully adapted for sea farming depending on the environmental control desired. Six zones are con-

sidered and analysed: (i) shore, (ii) intertidal, (iii) sublittoral, (iv) surface floating, (v) mid-water, and (vi) seabed, as shown in Fig. 2.

Prospective marine farmers wishing to utilise the coastal waters round the shore must also appreciate the legal formalities with regard to the necessary permissions required and conservation and legal size limit laws which apply to his selected species. In most cases the governments concerned control the area below high water mark, and the area required is leased to the farmer for a stipulated term; up to 25 years as in France. In many countries the rental for the lease is nominal, and is really a registration fee since the development of marine cultivation is encouraged.

Further the selection of a site on the coastline for a marine farm requires a critical assessment of (*a*) the supply of sea water, (*b*) the quality of the sea water (i.e., absence of pollution or toxic substances), and (*c*) the exposure of the site to storms. Each of these topics is discussed in detail in succes-

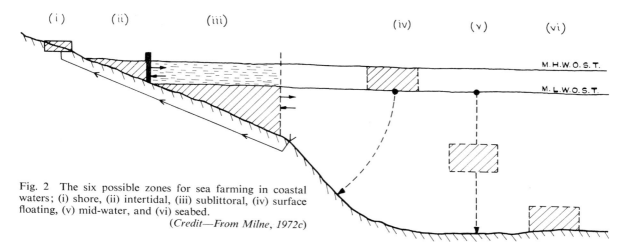

Fig. 2 The six possible zones for sea farming in coastal waters; (i) shore, (ii) intertidal, (iii) sublittoral, (iv) surface floating, (v) mid-water, and (vi) seabed.
(Credit—From Milne, 1972c)

sive chapters to ensure that the necessary preliminary site surveys are carried out before construction.

In addition to these surveys it is also essential to consider any modification the proposed farm may make on the environment. In the case of intertidal ponds, water exchange through sluices or the provision of pumps will be necessary. With netted enclosures in the sea the restriction of the mesh netting on the water currents must be assessed when considering the stocking density.

The effect of marine fouling growths on coastal structures and mesh must also be considered in calculating the design loads and estimating stocking densities, and the marine fouling of various materials is discussed in detail. Consideration also has to be given to the control of predators to prevent serious losses of stock, and various physical and chemical techniques for dealing with predators are mentioned. One of the techniques to combat the predation of sedentary organisms normally cultivated on the seabed, is to use off-bottom culture methods, which prevent attack by whelks, starfish and crabs.

The marine farmer has a certain amount of control over other predatory species, but the most difficult problem he may encounter is that of disease and parasites. These can, however, be prevented and controlled to a certain extent by careful management and husbandry, since some diseases are due to dietary deficiencies and physical and chemical factors in the environment. One of the major considerations in this field is that any transfers or introductions to new areas must be disease free to prevent indigenous stocks being wiped out. These two subjects of disease and parasites are considered to be outwith the scope of this book; Iversen (1968) has already given the prospective marine farmer

excellent advice on this topic, with reviews by Sinderman (1966, 1970a,b).

Many of the farming techniques described in succeeding chapters have been evolved and adapted to circumstances peculiar to that area (Prowse, 1963). Thus in considering the farming of the same species under two different environmental conditions this evolvement should be assessed, since it is seldom that one farming method can be transferred from country to country without alteration. An example of the difficulties involved in transfer is described in the research carried out in the United States of America into shrimp culture using the techniques previously developed in Japan. The basic principles are the same for all areas but methods must be adapted to the conditions peculiar to the new area under development.

The most outspoken objectors to the development of marine fish farms are the fishermen who earn their livelihood from hunting at sea. However, these fears are unfounded at present since the total production from farming is small in comparison to industrial fishing. The latest review of production from all attempts at cultivation of both fresh water and marine, both fish and shellfish for 1969 was 4×10^6 tonnes (FAO, 1970a), compared to $63 \cdot 1 \times 10^6$ tonnes for industrial fishing.

Several recent reviews of the status and potential of aquaculture have attempted to predict the development of this relatively new industry. The FAO Indicative World Plan for Agricultural Development (FAO, 1969) indicates an expansion factor of 5 by 1985 for both fresh and marine aquaculture, with a total annual production of 20×10^6 tonnes of high quality fish and shellfish for human consumption. Ryther and Bardach (1968), in their survey of world aquaculture, estimated a much more optimistic rate of increase to ten times by the year 2000, which would indicate a world production total of 40×10^6 tonnes.

To achieve such a rapid rate of expansion without expensive mistakes and failures it will be essential for the closest co-operation between the biologist and engineer in this bio-engineering venture, as discussed by Bell (1970). The biologist by studying the ecology of farmed species and their environmental tolerances for spawning and growth can thus advise the engineer of their requirements for sea farming. The engineer therefore has to arrive at economical design and construction methods to provide a suitable marine environment for the farmed species.

Finally, all concerned should appreciate that the construction of impoundments and enclosures in the sea will modify the coastal regime, and this must receive due consideration in any plans for large-scale commercial sea farming in our coastal waters.

Chapter 2
Fish and Shellfish Farming

The most important groups of marine organisms for sea farming are:

(i) molluscs (oysters, mussels and clams);
(ii) crustaceans (shrimps and lobsters); and
(iii) fish.

Both the molluscs and crustaceans are referred to as shellfish in this book to distinguish them from the vertebrate finfish. In shellfish farming the seabed or other artificial surfaces provide a habitat for the species. Whereas, in fish farming water volume containment is necessary either in a pond or net enclosure.

With a steady increase in fish and shellfish farming since the second world war numerous words have been coined to cover this topic, such as "mariculture", "marine cultivation", "marine farming", "sea farming", "fish culture", "aquaculture", and "aquiculture", all of which are similar in meaning and most confusing to the layman. The first four terms by the use of marine or sea imply the raising of organisms in a marine environment. The latter three are broader terms used to denote the raising of organisms in water whether fresh or marine. In this book which is devoted to the farming of fish and shellfish in the marine environment, of the first four terms, sea farming has been chosen as the most descriptive and least misunderstood term.

Similarly the words, "culture", "farm", "raise" and "rear", which are often used in the literature on sea farming are defined differently by various writers. The words, culture, raise and rear are used in this book in connection with the provision of good conditions for the growth and existence of marine organisms, whilst the term farm is used specifically for the commercial exploitation of the species either from hatchery or wild stock. Iversen (1968) chooses as his definition of sea farming:

"a means to promote or improve growth, and hence production, of marine and brackish water plants and animals for commercial use by protection and nurture on areas leased or owned."

Hence sea farming should only be applied to activities where there are concepts of husbandry and management in leased or owned areas where the owner has the sole right to harvest the stock.

Another term, that of brackish water, has now appeared, and this is a mixture of fresh and salt water usually under estuarine conditions. The salinity of sea water is expressed as the number of parts of salt contained in 1,000 parts of sea water (i.e., ‰). In the open sea the salinity is normally greater than 30‰. Due to rainfall and fresh water run-off from the land in some coastal areas, this salinity is naturally reduced, and brackish water can be considered in areas with a salinity of 15–30‰. Therefore brackish water farms tend to be in estuaries, and marine fish and shellfish farms in more open coastal waters, where the salinity is higher.

(I) MOLLUSCS

Oysters

There has always been a high demand for oysters throughout the world, and this is the most commonly cultivated marine organism with commercial farms in over fourteen major countries. The oyster is a sedentary mollusc with two hard shells, hinged together, and the opening and closing of the shell is operated by a strong muscle for breathing and feeding, Fig. 3.

The various oyster species farmed throughout the world fall into two groups, either flat or cup-shaped oysters. The flat oysters, so called because both shells are flat, belong to the genus *Ostrea*, of which there are two commercial species. The cup-shaped oysters possess a flat upper shell and lower cup-shaped or rounded shell, and belong to the genus *Crassostrea*, of which there are seven commercial species. Both are distributed over the world, mainly in warmer temperate waters. In addition

Fig. 3 Stick culture of oysters at Medoc near Arcachon. On right, Mon. le Dantec, Chef de la Station Expérimentale et de Recherches, Institut Scientifique et Technique des Pêches Maritimes, Arcachon. (*Credit—P. H. Milne*)

to the indigenous species, other varieties are often transplanted to replace or supplement existing stocks.

The European flat oyster, *O. edulis*, which has been cultivated since the days of the Romans is now farmed mainly in Britain, France (Brittany), Norway and Spain. The other flat oyster, *O. lurida*, or Olympia oyster, used to be farmed on the Pacific coast of America from southern Alaska to lower California. However, as the harvested oyster was very small, these native flat oysters have now been replaced in many areas by the imported Japanese oyster, *C. gigas*.

In Japan the farming of the Japanese oyster, *C. gigas*, has been developed successfully since the eighteenth century, with a wide export market for seed to Britain, Canada, the Philippines and the United States as mentioned above.

The main American oyster, *C. virginica*, is farmed extensively throughout the United States and Canada from the Atlantic coast of North America, southward around Florida, in the Gulf of Mexico, in California and Washington. The most specialised hatcheries and farming techniques have been developed in Long Island Sound.

The third important cup-like oyster, *C. angulata*, the Portuguese oyster, is found on the coast of Europe, and is mainly farmed in the Arcachon region of France and Britain.

Other traditional oyster species to be farmed are the Sydney rock oyster, *C. commercialis*, in Australia, the rock oyster, *C. glomerata*, in New Zealand, and the slipper oyster, *C. eradelie*, in the Philippines.

Recent developments in the cultivation of the mangrove oyster, *C. rhizophorae*, in Cuba and Venezuela, look promising for farming. In addition, Russia and South Africa have recently established oyster hatcheries and are experimenting with techniques for commercial application.

Mussels

Although there is a greater demand for mussels in Europe, they are not as profitable as oysters. However, in many cases they are easier to farm with less attention being required. The common edible mussel, *Mytilus edulis*, has a dark purple to black coloured shell, and attaches itself to rocks and collectors by means of a byssus thread, Fig. 4. It is cultivated by both hanging and bottom-culture techniques, often similar to the methods used for oysters. Spain is now the world's leading producer of mussels, marketing some 150,000 tonnes per annum. France, Holland and Italy are the other main European countries with large-scale mussel farms. The Spanish success with the floating raft culture of mussels has now led to developments in Germany, Ireland, Norway and Scotland to investigate the commercial prospects of many of their sea inlets for *M. edulis* farming.

The only other commercial mussel in Europe is *M. galloprovincialis*, the Mediterranean mussel farmed in Italy. Mussels are also cultivated in the Philippines, where the green mussel, *M. smaragdinus*, is farmed.

Several other species of mussel are also undergoing experimental cultivation techniques to determine their commercial prospects, and these are: *M. edulis planulatus* in Australia, *Perna canaliculus* in New Zealand, and *P. perna* in Venezuela.

Clams

Clams replace mussels in the United States as suitable sessile species for sea farming. The two

spawned artificially for seeding areas. In some places the hard clam is entitled differently. In New England it is called a "quahog" or "quahaug", and in the Middle Atlantic states is is termed a "hard clam", "hard-shelled clam", or "little-neck clam". Similarly the soft clam is sometimes called the "soft-shelled clam" or "long-necked clam" for differentiation.

Scallops

Although scallops are not yet farmed they are of considerable potential economic value, and research is currently under way in Japan on the Japanese scallop, *Patinopecten yessoensis*, and in Russia on *Mizuhopecten yessoensis* and *Spisula sachalinensis* to find suitable commercial techniques.

Abalone

The abalone like the scallop is also of potential economic value, and since there is a heavy demand for the northern Japanese abalone, *Haliotis discus,* cultivation attempts are in progress in Japan. At present the abalone are artificially spawned in hatcheries, and the seed sown in coastal areas for stocking. If the warm water effluent from coastal power stations was utilised this could increase its commercial prospects.

(II) CRUSTACEANS

Shrimps

Of the crustaceans suitable for sea farming, the shrimp is the most important and is extensively farmed over the world. Since shrimps grow rapidly within impoundments and are in great demand they are ideal for intensive cultivation, Fig. 5. In some countries they are caught in the beach zone of estuaries and placed in ponds for raising. In other countries ponds are constructed in areas where the shrimp are indigenous so that the tidal currents carry the young larvae into the pond. Sea farming by this technique relies on an adequate supply of shrimp, but is limited by the presence of larval forms of predatory species also carried into the ponds at the same time. The various shrimp species farmed throughout the world fall into two groups, *Penaeus* and *Metapenaeus*.

Singapore has the largest number of indigenous commercially attractive shrimps, and the sea farming here relies on the young postlarval shrimp being carried into the ponds by the tidal currents. The main species harvested from the ponds in Singapore are *P. indicus, P. merguiensis, P. monodon, P. semisulcatus, M. brevicornis, M. burken-*

Fig. 4   Rope cultivation of mussels suspended from a floating raft. This rope shows the growth of mussels after six months in Linne Mhuirich, Argyll in Scotland.

(Credit—This photograph is Crown Copyright reserved and is reproduced by permission of the Controller of H.M. Stationery Office)

species are the hard clam, *Mercenaria mercenaria,* indigenous along the Atlantic coast from Maine to Florida, and the soft clam, *Mya arenaria,* found from Labrador to North Carolina. These two clams are very suitable for cultivation since they can be

Fig. 5 Part of a catch of shrimp taken in the Arabian Sea where an expanding industry is developing. (*Credit*—FAO)

roadi, *M. ensis* and *M. mastersii*. The farming of *P. monodon* with milkfish is also practised along similar lines in the Philippines.

Considerable attention has been paid in Japan in the last century to the commercial farming of the Japanese shrimp, *P. japonicus*, following the extensive research of Dr Fujinaga (Hudinaga) as described in Chapter 6. The techniques developed in Japan have now received considerable attention, and modifications in these methods are now under study in the United States, Indonesia, Britain and Australia.

Many of the United States government, university, and industrial laboratories have for some time been trying to adapt the Japanese techniques of shrimp culture to their own native species, the pink shrimp, *P. duorarum*, the brown shrimp, *P. aztecus*, and the white shrimp, *P. setiferus*. Most of this work is still research, but recently 1,000 hectares of West Bay, Florida were leased for large-scale commercial farming.

Recent successful shrimp research in Indonesia has encouraged the construction of a shrimp propagation centre for the commercial farming of *P. merguiensis*, *P. monodon*, *M. brevicornis* and *M. monoceros*. A start has also been made in Australia into feasibility studies for the farming of the king prawn, *P. plebejus*, the greasyback, *M. bennettae*, and the school prawn, *M. macleayi*, which are all of commercial potential.

The first experimental shrimp farm to be established in Britain was constructed adjacent to Hinkley Point Power Station to utilise the warm water effluent. Earlier research on the English prawn, *Palaemon serratus*, carried out by the Ministry of Agriculture, Fisheries and Food, had demonstrated methods of obtaining larvae from mature adults in the laboratory, and these techniques have been adopted and developed for large-scale farming at Hinkley Point.

Lobsters

At the beginning of the twentieth century considerable attention was focused on both the American lobster, *Homarus americanus*, and the European lobster, *H. vulgaris*, and several large-scale lobster hatcheries were built for the controlled production of lobster larvae, which were then released in the sea in the hope that it would boost the commercial catch. Unfortunately these attempts were unsuccessful due to the high mortality in larval form.

Although lobsters are a species which can be grown to marketable size under controlled conditions the five years required to reach this stage has deterred their exploitation. However, recent experiments using heated water in America have shown a reduction in the time required to raise marketable lobsters and this has increased their viability for sea farming. Another enterprise in Scotland has successfully shown that lobsters can be held in captivity in seabed cages for commercial exploitation, Fig. 6.

Turtles

Although turtles are reptiles and not crustacea or fish, they have been included in this section since like the lobster they have a carapace.

The meat and soup obtained from the green turtle, *Chalonia mydias*, have long been desirable gourmet

Fig. 6 These European lobsters were caught in a lobster pot for transfer to seabed cages for fattening at Loch Inchard, Sutherland in Scotland. (*Credit—P. H. Milne*)

foods, but unfortunately due to their over-exploitation by man their natural numbers have been considerably reduced. To reverse this trend, and produce a regular supply of turtle meat, attempts are now being made to farm sea turtles. The first large-scale commercial farm has been constructed on Grand Cayman Island in the British West Indies. Similar attempts to farm turtles are also under way in Torres Strait, Australia.

(III) FISH

Mullet

Mullet farming has a long history, which can be traced back to the use of brackish water impoundments in early Roman times. The most common species currently farmed is the grey or striped mullet, *Mugil cephalus*. It is frequently farmed in conjunction with other fresh and brackish water fish, in India and Israel. The exclusive farming of mullet in marine waters is largely confined to France and Hawaii. In France the ponds are stocked by inducing the indigenous mullet to swim into the ponds with no artificial stocking. Whereas, in Hawaii young mullet fry are netted in the estuaries and transferred to coastal enclosures for growing and fattening.

Yellow-tail

The farming of yellow-tail, *Seriola quinqueradiata*, in Japanese coastal waters is a major industry and accounts for 98·6% of the total production of sea farming in Japan. Three techniques are applied for

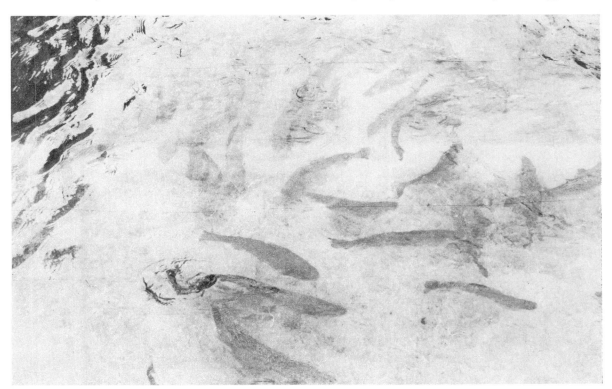

Fig. 7 Rainbow trout culture in a concrete raceway. These 225 gm fish are an ideal table weight. (*Credit—P. H. Milne*)

Fig. 8 Atlantic salmon harvest from a salt water enclosure in Norway. (*Credit—by courtesy of A/S Mowi*)

Fig. 9 Selection of five-month old pond-reared pompano from Tampa Bay, Florida.
(*Credit—U.S. Department of Commerce—National Marine Fisheries Service Biological Laboratory*)

the farming of yellow-tail, the most common being the use of floating net cages followed by coastal ponds and netted enclosures.

Rainbow Trout

Although the American rainbow trout, *Salmo gairdnerii*, hatches in fresh water, it may be acclimatised to sea water for growing and fattening in commercial farms, Fig. 7. The Norwegians led the developments in this field and have now established a lead in the commercial development of "marine" rainbow trout. Scotland, with similar water conditions to Norway, has also embarked on commercial rainbow trout farming in the sea. Australia has also followed suit, having first established a fresh water farm, they are now developing marine techniques for sea farming.

Salmon

The salmon, like the rainbow trout, hatches in fresh water before returning to the sea. Several hatcheries have been built over the years to replace the lost natural spawning conditions of salmon rivers which have been dammed for hydro-electric development (Mills, 1971). However it is only comparatively recently that with dwindling salmon

catches at sea, interest has focused on the sea farming of salmon.

Sea farms have now been established for farming the Atlantic salmon, *Salmo salar*, in Norway and Scotland, Fig. 8. The Pacific salmon, *Oncorhynchus* sp., has received attention in the United States, and the various species now under development are the chum, *O. keta*, coho, *O. kisutch*, pink, *O. gorbuscha*, and chinook, *O. tschawytscha*.

Pompano

Several attempts have already been made to farm the common pompano, *Trichinotus carolinus*, Fig. 9, but so far without success due to lack of knowledge of their biology and ecology. Since pompano command a high price several research programmes are now under way to develop reliable commercial farming techniques.

Plaice

The plaice, *Pleuronectes platessa*, is one of the flat-fish found round the coast of Britain and in the North Sea. Research on this species, carried out in the late nineteenth and early twentieth century, showed that trial transplants to the Dogger Bank improved the general fishing in the area, but since

Fig. 10 Plaice culture in Britain. A fine specimen of a two-year-old plaice reared at Ardtoe, Argyll in Scotland in an experimental floating cage.

(Credit—Michael Wood, by courtesy of the White Fish Authority)

fishing boats from all nations had access to the stock these were discontinued.

Further experiments in Scotland during the second world war, when plaice and flounders were raised in a fertilised sea loch showed promise, but the development of retention techniques was not included and the fish subsequently migrated out of the area under study.

Recent research carried out in Britain by the Ministry of Agriculture, Fisheries and Food, and the White Fish Authority into the commercial development of plaice in ponds and floating cages appears promising, Fig. 10.

Sole

The sole, *Solea solea*, is another flatfish found in British waters, but it apparently requires warmer water than plaice. Research by the White Fish Authority has therefore been restricted to investigating the possibility of utilising the waste heat from coastal power stations for development.

Tuna

The possibility of raising bluefin tuna in floating net enclosures is at present under study in Japan using wild stock fry caught inshore. It is also hoped to develop hatchery spawning techniques in the future.

Sea Bream

The sea bream is a popular fish in France, and a recent farm has been constructed using old salt ponds in the intertidal zone for its commercial exploitation.

Chapter 3
Legal Aspects of Sea Farming

It has often been said that area for area the sea is more productive than arable agricultural land. However, in order to develop the farming of the species selected it is essential to be able to exercise control over the area concerned. On private land for the development of hatcheries or shore tanks this is possible. In the sea numerous legal regulations restrict the marine farmer, not only with regard to obtaining wild stock for growing to maturity, but also in the marketing of species in the close season, and in the regulations governing maritime construction.

The prospective marine farmer must therefore, before establishing a farm, consult the local state or federal conservation agencies, development boards, navigational authorities and government fishery authorities for the current regulations, restrictions and developments affecting his aspect of marine cultivation.

Conservation Laws

Most of the developing countries throughout the world have passed regulations preventing the over-exploitation of the natural fishing stocks. These laws are implemented by several restrictions (Iversen, 1968)

    (i) on the minimum size limits of fish sold;
    (ii) on the total quota of fish caught;
    (iii) by closed seasons of the year when fishing is prohibited; and
    (iv) on the types of fishing gear permitted.

Since these laws were designed for the industrial fishing industry it might appear that they would not affect the marine farmer, but in fact some regulations govern both wild and cultivated species, and care must be taken to ensure which regulations apply in different countries.

Minimum Size Limits

The conservation laws mentioned above impose minimum size limits in order to protect spawning grounds and allow young fish, shellfish, and crustacea to grow to maturity to preserve the existing fishery. It is thus illegal for anybody to possess or sell fish, shellfish or crustacea below these minimum marketable size limits. However, the marine farmer at the outset has to obtain wild stock fry for his farm, and unless he wishes to maintain a large hatchery with mature fish for spawning, it is illegal for him to catch fry which are by definition undersized fish for farming. He therefore requires a special permit to allow him to catch fry. This in general is not unreasonably refused since the natural predation of these small fry is normally so severe that only a small percentage reach marketable size, Fig. 11.

Fig. 11 Beach seining for fry, in this case pompano, on the Gulf beaches of Florida's west coast.

(*Credit—U.S. Department of Commerce—National Marine Fisheries Service Biological Laboratory*)

Closed Seasons

The original intention of the conservation laws in creating closed seasons, and restricting the fishing for several species, was to protect the species during the breeding season. In some cases the closed seasons were designed to ensure that the species was not over-fished. Again this would not appear to affect the marine farmer, but often these laws stipulate that the species must not be sold during the closed season either, and this is to the detriment of the marine farmer, who could obviously command a high price for his stock during this period.

An example of this is the oyster farmers in the State of Georgia (Iversen, 1968) who can only sell their oysters during the months when they can be legally caught from public grounds. This means the farmers have to tend their stocks during this period even although they may be of marketable size.

Fortunately, however, some governmental fishery agencies appreciate this problem, and allow, under special licence, the selling of stock during the closed season. For example, in Hawaii (Iversen, 1968), pond-raised mullet can be sold during the closed fishing season from January to March.

Ownership of Foreshore and Seabed

The prospective marine farmer wishing to acquire a certain stretch of foreshore, intertidal beach or seabed for marine cultivation is beset with many laws and regulations, which in some cases restrict development. The following is a general resumé of the position, but the relevant government department should be consulted for information.

If the proposed site is on the shore, or is attached to the shore above high water in any way, for example, moorings or access catwalks, the permission of the land owner is obviously required for access and construction. In favourable circumstances it may be possible to purchase the ground; on the other hand the solution may be to rent the ground on a long term lease.

The intertidal beach section normally belongs to the state and comes under their control. Unfortunately in some cases the local laws prohibit leasing in areas suitable for sea farming. For example in Maine, New Hampshire and Massachusetts in the United States the laws give free access to all waters and beaches, thus hindering the development of new techniques, since wild shellfish have no owners, and there is no encouragement to improve the fishery (Iversen, 1968). A very good review of the controls on ownership of intertidal areas in connection with shellfish has recently been published by Nowak (1970), and traces the history of the laws and regulations. In many countries, however, it is possible to lease sections of the beach, with sole control of the cultivation of the marine organisms on site. The leasing of areas for private oyster farming, as in the State of Washington, where the marine farmer is unencumbered by fishery laws allows full management control and development. Ryther and Bardach (1968) produced some very illuminating figures illustrating this

difference. In general the leased oyster beds in the United States produce an average of 1·50 tonnes per hectare per year, whilst the public grounds only produce 0·075 tonnes per hectare per year.

In general it is advisable to obtain a long lease of the area concerned to ensure that the money spent on preparing it for marine cultivation is not lost at the end of the lease. In France the areas used for oyster and mussel culture are leased from the government for 25 years. In Australia the oyster beds are leased from the government for 15 years. In Spain the government own and control the bay areas used for raft mussel culture. These leases are for 10 years, and are renewable, but may be terminated by the government without notice. However, with the raft system, this is not too critical since at least the raft can be towed to another site. In the Philippines both the oyster and mussel culture areas come under municipal governmental control. In some towns the oyster growers are registered and are assigned to specific areas, but in other towns there are no formal arrangements, and the individuals assume squatters rights (Ryther and Bardach, 1968). Of the shrimp ponds also in the Philippines half are privately owned and the other half leased from the government. Most of the shrimp ponds in Singapore have been developed by the local government agencies and the ponds are leased. When the last four ponds were opened up in a new area there were 1,000 applicants for the leases, which indicates the interest in marine cultivation there.

In some countries laws enable sport fishermen access to impoundments over a certain area in size, whether man-made or natural (Iversen, 1968). This law may benefit sport fishery but it acts to the detriment of marine farming where complete control of access is essential.

Maritime Waterways

Where the construction of a marine farm impinges on a maritime waterway, permission will have to be obtained from the navigation authorities. If the intended construction is near a busy waterway, or near a navigation channel, navigation lights may be necessary at night, with the provision of radar reflectors. However, the proximity of a commercial waterway should be viewed with caution because of the likelihood of pollution as discussed in Chapter 13.

It is more likely, however, that the structure may affect bodies running commercial-type tourist vessels, or local supply boats, and such organisations would have to be consulted. If an intertidal pond is to be constructed, the sealing-off of access to the sea from certain shores may require some compensation or replacement slipways to ensure the goodwill of those who have previously enjoyed freedom of access. In the case of a sublittoral enclosure with a net barrier across the waterway, a boat lock like those developed in Japan will be necessary, and these are described in Chapter 8, which deals with sublittoral enclosures in detail.

2
PRELIMINARY SURVEYS

Chapter 4
Site Selection & Analysis

Prior to the establishment of a commercial marine farming venture on any site it is essential first to carry out a careful background analysis of the hydrology and hydrography of the site to ensure satisfactory sea water supplies for the farmed species. Secondly it is necessary to consider the exposure of the site to winds, waves and tidal currents. An assessment has also to be made of any changes in the environment due to construction, and this is discussed in Chapter 11 in relation to fish farms described in Chapters 6 to 10.

Site Selection

Various techniques are of course available for the retention of the farmed species, and it might be advantageous to discuss first the possible variety of zones suitable for marine farming, moving progressively offshore (Milne, 1969a, 1972c), Fig. 2.

(i) shore,
(ii) intertidal,
(iii) sublittoral,
(iv) surface floating,
(v) mid-water, and
(vi) seabed.

If full control of the environment is required, as with a hatchery for spawning and larval production, the most suitable facilities are tanks or ponds on the shore or in the intertidal zone. During rearing to marketable size, if either the pond bottom requires treatment, or if it is desired to drain the pond completely for harvesting it is also essential to locate the enclosure either on the shore or in the intertidal zone. Shore construction has the added necessity of pumps to supply the farm with water for circulation and replenishment. Intertidal construction, if the area is well selected, should only require sluice gates for control, but often it is better to instal pumping plant to be able to increase the circulation.

For sessile organisms like molluscs where enclosure containment is not necessary, either the

intertidal, or seabed zone can be used to advantage for layering and transplanting. However, in some cases the seabed is not suitable if it is very soft and silty, and the skilful use of sublittoral or floating off-bottom techniques may increase the farming opportunities of an area.

Where control of the environment is not necessary, the tidal range may be utilised to provide the circulation and the location used depends on the access required with a choice of sublittoral enclosures, floating and seabed cages.

Each of the locations listed may possibly be used for marine cultivation, but often the biological requirements of the farmed species determine the most suitable method in a certain area due to variations in tidal range and the air and water temperatures.

Construction of shore facilities is of course the simplest to plan since the only tidal working is associated with the sea water inlet and outlet. The hydrography of the coastal waters must thus be studied to ascertain seasonal changes in water temperature, salinity and dissolved oxygen content. If large ponds are to be excavated, which have a natural catchment from the land, the intensity of rainfall must be determined, and provision made for its diversion, or the salinity of the pond will become lowered after heavy rainfall. In brackish water ponds this may be quite acceptable but in marine ponds this reduction in salinity may impair development of the fish.

The intertidal zone is traditionally used in many countries for mollusc culture and requires regular monitoring of the hydrographic environment offshore and in the intertidal zone at high water to ensure adequate growing conditions. If natural spawning conditions exist in the area, and are to be used for the initial stock, preliminary trials should be carried out with spat collectors to ensure an adequate settlement for commercial exploitation.

The building of ponds in the intertidal zone has been carried out in mangrove swamps in the Far

Fig. 12   Fresh water diversion from upper reaches of inlet at Ardtoe, Argyll in Scotland. In foreground 0·8 ha fresh water pond. In middle, 1·2 ha salt water pond, with Atlantic in background.   (*Credit—P. H. Milne*)

East for brackish water fish culture for several centuries. However, these same conditions can only be conducive to marine fish farming if the salinity can be controlled. As mentioned for ponds onshore the rainfall from the surrounding catchment must be assessed since it is essential that this fresh water does not become impounded in the pond (Milne, 1971c). If an intertidal area is sealed off, the fresh

water draining into that area may need to be diverted, Fig. 12.

The operational level of ponds in the intertidal zone is also most important. The ideal site for such a pond is shown in Fig. 13 (Milne, 1972c) where the main criterion is that the pond should have as large a volume between high and low water neap tides as possible. If the pond level is maintained at the

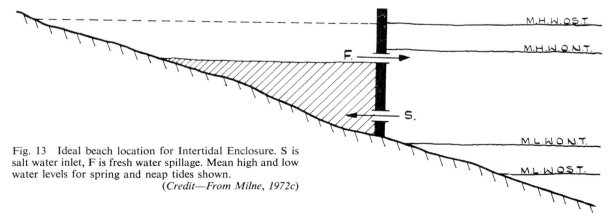

Fig. 13   Ideal beach location for Intertidal Enclosure. S is salt water inlet, F is fresh water spillage. Mean high and low water levels for spring and neap tides shown.
(*Credit—From Milne, 1972c*)

M.H.W.O.S.T.

M.H.W.O.N.T.

F.

S.

M.L.W.O.N.T.

M.L.W.O.S.T.

Fig. 14 Regular maintenance of sublittoral net enclosures by divers. At Megishima in Japan seven divers are employed out of a staff of 25 to clean the 1,370 m of net barrier. (*Credit—by courtesy of the White Fish Authority*)

high water neap tide level, water circulation and replenishment can be effected at any time in the spring-neap tidal cycle. Also if the bottom is chosen above the low water neap tide mark, the pond may also be drained for harvesting or maintenance at any given low water. The benefit of onshore or intertidal ponds is the ability to vary the salinity as required.

If, however, it is not necessary to have a specific salinity, and the ambient salinity of the region under consideration fluctuates only within the limits of the proposed species, it is possible to think in terms of a net enclosure. This may take the form of a rigid net barrier to seal off a bay (Milne, 1970b,e), or may be a cage either floating or moored to the seabed.

Sublittoral enclosures have the advantage of being relatively inexpensive methods of enclosure, but with additional maintenance costs to keep the nets clean to preserve the water circulation, Fig. 14. The main criterion for the selection of sublittoral sites is that the variation between high and low

water should be as small as possible, bearing in mind that there should also be adequate recirculation within the enclosure at neap tides. The reason for this is that it is the low water volume which determines the density of stocking of the enclosure and this relationship is discussed later in Chapter 11 with reference to several Japanese net enclosures.

The choice of floating structures, either rafts or cages, simplifies the construction and husbandry of the species. Also in unfavourable weather or unsuitable environmental conditions, it is always possible to tow floating units to safety. Floating rafts for mollusc cultivation, where the organisms feed off the water-borne plankton, only require to be visited for maintenance and inspection, Fig. 15. Floating cages for marine fish farming, however, suffer from the disadvantage of requiring regular access for feeding, Fig. 16. This is often simplified by grouping several cages together, Fig. 17, as discussed in Chapter 10. Another simplification is the use of amphibious vehicles for the servicing and feeding.

Fig. 15   Floating raft culture of molluscs, mussels and oysters, in Spain. These rafts are the hulls of old ships with wooden frameworks suspended over the side, held fast by backstays to the mast. (*Credit—P. H. Milne*)

The use of the seabed below the low water spring tide mark has been fairly extensive in the past for mechanised mollusc culture. However, if reasonable harvests are to be maintained it is essential for regular SCUBA diving inspections to restrict starfish and crab invasions. If cages for crustacea are moored to the seabed, divers are also required for regular feeding and maintenance. Seabed cages, however, can be built very cheaply

Fig. 16   Floating cage for plaice culture moored alongside landing stage at Ardtoe, Argyll in Scotland.
(*Credit—P. H. Milne*)

as they are outwith the turbulent air and sea interface with a consequent reduction in the design considerations.

Hydrographic surveys of the area under consideration, as well as studying the water currents, temperature, salinity and dissolved oxygen content, must also ensure that the area is free from the pollution of industrial wastes and sewage discharges and is clean from a sanitary viewpoint. It is advisable that hydrographic observations are obtained over an annual cycle at any specific site for development to ensure satisfactory conditions throughout the year.

Site Exposure

Once an area has been chosen on the merits of providing the biological requirements of the species to be farmed the next consideration is the exposure of the area to winds and waves, Fig. 1. Under exceptional circumstances it may be essential to design the structure to withstand hurricanes and typhoons.

Any structures standing above the water level either in the sea or on the shore will be subjected to dynamic wind forces, and the "probability" of occurrence of these winds must be analysed from the extreme meteorological conditions pertaining

Fig. 17 Groups of floating cages connected by spider framework and central platform. These cages are used for farming rainbow trout and Atlantic salmon in Loch Ailort, Inverness-shire in Scotland. The hatchery and shore tanks are in the background, and an amphibious vehicle is used for access. (*Credit—by courtesy of Marine Harvest Ltd*)

to that area. Normally the Meteorological Office of the country concerned will have kept wind records and it is necessary to determine both the mean hourly wind speeds and the maximum gust speeds occurring in the vicinity of the site. Any construction works above low water must therefore be capable of withstanding the maximum gust speeds for the duration or life of the farm. The underwater sections of the farm facilities, however, only need to be constructed to withstand the wave forces, which are induced by the wind blowing over a period of time and the mean hourly wind speeds are used in this determination.

For solid structures the determination of design forces can be calculated quite easily, but for mesh retaining structures the evaluation of design forces is more complex. A method for their ready calculation has been evolved (Milne, 1970e) and is presented in Appendix 1 for those requiring to design such a structure.

In addition to considering the wind forces on coastal structures, an assessment must also be made of the variations in sea water level due to meteorological disturbances. Combined wind and barometric pressure effects often give rise to an increase in the water level above the maximum spring

tide level, called a storm surge, and these often sweep inland on very exposed coasts. The main countries experiencing storm surges normally have meteorological warning services for such occurrences, such as in the Gulf coast of America (Bretschneider, 1967) and the east coast of Britain. However, these surges can take place on any exposed coastline, and such a prominent one occurred on the west coast of Scotland in 1968 (Milne, 1971b).

If either sublittoral enclosures or floating structures are contemplated the heights of the highest probable waves occurring there must be ascertained, in order to calculate the wave forces. Such coastal structures are normally designed to withstand the highest single wave expected, and this design wave height is calculated from:

  (i) the characteristics of the wind field;
 (ii) the direction and speed of the wind field;
(iii) the fetch length (open water distance to nearest land); and
 (iv) the water depth variations along the fetch.

Thus any possible enclosure or raft site has to be investigated uniquely and the design wave for the site will depend on the geographical location, and topographical features in addition to the wind speed. A discussion on the analysis of the effective fetch for a site and the determination of wave heights and forces from this is included in Appendix 2 for reference (Milne, 1970e).

Last, but by no means least, the effect of tidal currents on the structure must also be calculated. Any enclosure or raft is dependent on the strength of the tidal currents for water circulation and dissolved oxygen replenishment, and any reduction in these, due either to bad design or fouling of the mesh netting, may cause lack of circulation. The strength and direction of these tidal currents should be ascertained in the hydrographic survey to ensure the correct design and orientation of the units. Since mesh nets in the sea will collect a certain amount of marine growth (as discussed in materials selection in the next chapter), it is therefore essential to calculate the design forces on the units with fouled nets, or there is a risk of the nets being carried away by the waterflow, as did happen at Matsumigauru in Japan, as described in Chapter 8. Curves for the determination of tidal current forces on fouled mesh panels of various materials are presented in Appendix 3, showing how to calculate the necessary design forces (Milne, 1970e).

Chapter 5
Selection of Materials

The selection of materials for the enclosure and retention of the cultivated species is as important as the selection of the site.

In any marine farming operation the use of screens and nets are obligatory to prevent: (i) the admission of predatory species; and (ii) the loss of the stock through sluice gates and outlets.

The choice of materials for structural work either in the intertidal, sublittoral or surface floating zone, is also critical in that:

(*a*) the structure must have a reasonable life expectancy with materials that do not corrode or decay in sea water;

and

(*b*) the chosen material should be relatively resistant to marine fouling growths, since these can impose an additional load on the structure. Alternatively the structure should be capable of being cleaned easily, but this, however, increases maintenance costs.

Mesh Fabric Review

Numerous materials of natural, polymeric and metallic composition are capable of being made into mesh fabric, which can be used in marine fish farming for screening purposes. The best fabric for use at any one site depends not only on the physical properties of the fabric but also on its resistance to fouling and deterioration in a marine environment (Milne, 1970b), since it is essential that the mesh remain free from marine growths to allow free circulation of sea water. The type of fabric construction and its effect on the handling and storage of mesh netting is also discussed since this is most important when considering the installation of mesh screening and its subsequent maintenance.

Of the various types of fabric used, hemp and cotton are natural fibres and nylon, Terylene, Ulstron and Courlene are man-made polymeric fibres. All of these fibres can be made into twines suitable for the manufacture of mesh netting of the conventional knotted mesh type. They are all used for standard fish nets which can be made to any size with a diamond mesh, but can be obtained in a square mesh if desired. These nets are very easily handled and stored as they are very flexible and can be bundled up when not in use.

A plastic extruded mesh which is being used extensively outdoors for fences and trellis work is Netlon. It is a semi-rigid square mesh which retains its mesh size, but is flexible enough to be rolled up for storage purposes. A survey was recently carried out into its particular use in oyster and mussel culture (Nortene, 1969), when Netlon's potential use for spat collectors was examined.

Another polymeric fabric used for fences is Plastabond, a polyvinyl chloride covered galvanised steel wire, which is normally woven into chain link diamond mesh. It can be supplied in up to 2·75 m widths in 20 or 45 m long rolls, so can be easily handled and stored.

Metallic fabrics, such as galvanised steel, aluminium, stainless steel, brass, copper, cupro-nickel and nickel, can all be supplied in cold drawn wire form. These wires can be used to make either a chain link diamond mesh, which can be rolled for easy handling and storage, or the wires can be welded together to form square mesh. With galvanised steel it is better to weld the mild steel wires together before galvanising. Some of the finer metallic wire meshes can be rolled, but the heavier gauges are supplied in sheet form, which means they have to be stacked, and the sheet size is dependent on the handling weight and limitations of transport. The metallic fabrics can also be supplied in expanded metal sheets with diagonal apertures and can be obtained either honeycombed or flattened. It is not possible to roll this type of fabric construction and sheets of this material must be small enough for handling and stacking for storage.

Marine Fouling

The constant immersion of mesh netting in sea

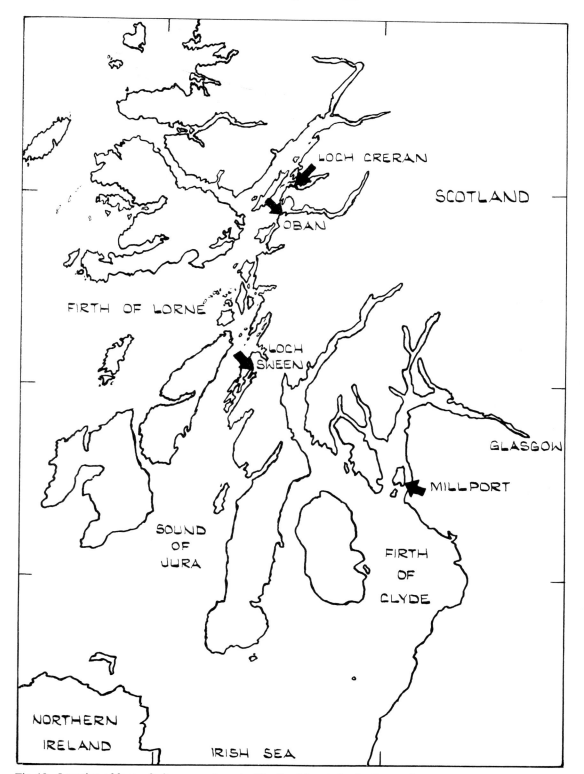

Fig. 18   Location of four raft sites on west coast of Scotland for marine fouling studies on test panels of fish netting.
(*Credit—From Milne and Powell, 1972*)

water presents marine fouling problems not normally occurring in the realm of the fishing industry. Nets used for hunting fish are usually cast for only a short time, and any settling organisms die on exposure when the net is recovered and stacked on board.

When the British Natural Environment Research Council sponsored research by the Department of Civil Engineering at the University of Strathclyde into the design and construction of sublittoral enclosures for marine fish farming (Milne, 1970b,e), it was concluded that initial fish cages and enclosures would use mesh netting for the retention of the fish. However, very little detailed information was available on the marine fouling of mesh netting permanently immersed in the sea.

It was therefore necessary to study the marine fouling of various submerged netting fabrics at different locations in order to select the most suitable one for use in the initial marine fish farming experiments. This research was carried out in the period 1967–71 by the University of Strathclyde in conjunction with the Scottish Marine Biological Association (S.M.B.A.) at four sites on the west coast of Scotland, Fig. 18 (Milne and Powell, 1972).

From fishing experience, it was known that hemp and cotton were very prone to biological attack, but it was thought that modern synthetic fibre fishing nets were more resistant to marine fouling. Ten panels of mesh netting fabric which included samples of synthetic fibre, plastic and metallic materials were chosen for this research. As a raft was

immediately available for the immersion of panels at Millport, alongside the S.M.B.A. Laboratory in the Firth of Clyde, work commenced there in July 1967.

In these initial tests samples of standard 25 mm fishing nets of nylon, Ulstron, Courlene, polythene and cupra-proofed polythene were each immersed vertically on individual galvanised steel frames, 0·66 m square, with the top bar approximately 0·3 m below the surface. More rigid polymeric fabrics were also immersed, such as Netlon and chain-link Plastabond. Galvanised steel was the only metallic fabric readily available in a 25 mm mesh form, either in chain-link or square weld-mesh, so the more expensive metallic fabrics were not considered for the initial survey. These panels were subsequently examined, photographed and weighed in air (after draining) every two months, to ascertain the settlement sequence of any fouling organisms, following the recommendations of Brandt and Carrothers (1964).

The first ten panels were immersed at Millport in July 1967 and after one month the panels were lightly covered with a brown filmy growth of algae (mainly *Ectocarpus* sp.), with a few tufts of *Tubularia larynx*. This slime film and its subsequent role in marine fouling was expected following the recent review of this topic by Horbund and Freiberger (1970). After two months, in September 1967, the panels had become heavily fouled, mainly with mussels, *Mytilis edulis*, with an average size of 7–8 mm. There was also a light settlement of *Tubularia larynx*, and a few *Facelina auriculata* which feed on

Fig. 19 Marine fouling test panels at Millport, Firth of Clyde in Scotland after two months immersion—net identification: 1—Nylon, 2—Ulstron, 3—Courlene, 4—Polythene, 5—Polythene (cupra-proofed), 6—Plastabond, 7—Galvanised chain-link, 8—Plastabond, 9—Netlon, and 10—Galvanised weldmesh. (*Credit—From Milne and Powell, 1972*)

the *Tubularia*. The galvanised panels had the least settlements of mussels but more algae and *Tubularia*, Fig. 19.

After four months, in November 1967, the settlement of mussels on the synthetic mesh panels had completely obscured the netting. However, the galvanised steel panels had significantly fewer mussels than the other panels, but more algae were present, mainly *Cladophora* sp., *Ulva lactuca*, *Ceramium rubrum*, *Ectocarpus* sp., and old basal parts of *Tubularia*. After six months, in January 1968, the mussels, average size 1·3 cm, were dominant on all the panels with a reduction in algae.

The results of the marine fouling observations at Millport were very useful and gave a good indication of the relative marine fouling characteristics in the Firth of Clyde, but could not be considered typical for the west coast of Scotland. It was then decided to extend this work and to carry out similar experiments on the west coast of Scotland, to determine the local fouling characteristics and sequences, and variations in marine fouling from place to place, and year to year.

Three more rafts were installed in 1968 at Loch Sween, Loch Creran and Dunstaffnage Bay, the latter alongside the new S.M.B.A. Laboratory, Fig. 18. The first two sea lochs, Loch Sween and Loch Creran, were chosen since they appeared to offer promising situations for marine farming. This was subsequently confirmed by the establishment of a mussel farming venture at Loch Sween and an oyster hatchery at Loch Creran (Milne, 1970c). Marine fouling studies were started at these three new rafts in the spring of 1968 in addition to Millport, and a further series of nets was added to the Millport raft to give a comparison with the previous year's fouling (Milne and Powell, 1972).

The marine fouling sequence at Loch Creran and Dunstaffnage Bay was similar to that at Millport, with the initial fouling consisting of brown filmy growths of algae up to July, and a subsequent mussel settlement by September, which grew to block some 50% of the synthetic fibre meshes by November. As before, the galvanised steel samples had lighter settlements of mussels.

The Loch Sween raft sited at Faery Isles, Fig. 18,

Fig. 20 Marine fouling test panels at Loch Sween in Scotland after eight months immersion. Net identification as in Fig. 19. (*Credit—From Milne, 1970e*)

Fig. 21   9 m long experimental "H"-frame netting barrier at Faery Isles, Loch Sween in Scotland. This was built to study civil engineering construction methods and also the marine fouling of large netting panels 4·5 m × 1·5 m.

(Credit—P. H. Milne)

showed the same initial settlement of algae by July, but this was followed by a massive settlement of ascidians, *Ascidiella aspersa* and *Ciona intestinalis*, which virtually obliterated all the synthetic mesh and polymeric fabric apertures. By November, very little mesh was still visible, the upper halves of the nets being colonised by mussels which had settled on top of the ascidians, Fig. 20. The mussels eventually overwhelmed the ascidians and were in overall control of the whole panels by the following January. In contrast, very few ascidians or mussels settled on the galvanised steel samples, and those only at the corner of the meshes.

At the same time as installing the new rafts, it was felt that studies of larger panels of mesh would be desirable, and if possible these should be fixed to span the intertidal and sublittoral zones. To assist with this a 9 m long experimental "H"-frame netting barrier, 5 m high was constructed by the University at Faery Isles, Loch Sween in 1968, Fig. 21, to study civil engineering construction methods and materials for marine farming in tidal waters (Milne, 1970e), as discussed in Chapter 8. Six types of mesh netting fabric, in panels 4·5 × 1·5 m were hung on the framework from above the

high water spring tide level down to the seabed, 2 m deep at low water spring tide, thus giving information on both the intertidal and sublittoral fouling.

Although galvanised weldmesh gave the best results in these tests it appears that the mesh will have to be replaced every three to five years (Butler and Ison, 1966) depending on the exposure and water currents. Enquiries into the possibility of using the other metallic fabrics, mentioned earlier, produced samples of 90/10 cupro-nickel chain link fencing (Anon., 1969) and Mn/Al square mesh for testing. In addition a new polymeric material called Parafil was developed especially for moorings with an anti-fouling coating and samples were obtained for testing. Three samples of each of these three materials, together with three samples of Courlene and galvanised weldmesh were subjected to comparative fouling tests at Loch Sween in 1970–71.

One panel of each material (i.e., five panels) were suspended from the Faery Isles raft in its sheltered location. Another five panels were suspended at the surface in open water, and the third five panels were submerged at a depth of 10 m at low water, both in Loch Sween.

The results of this test were most illuminating. The panels at Faery Isles developed the same fouling characteristics as before; galvanised weldmesh with little fouling, the others having heavy settlements of ascidians and mussels. The surface panels in Loch Sween showed no trace of ascidians but had a heavy settlement of mussels on all but the Cu/Ni alloy and galvanised weldmesh panels. The bottom panels at 10 m had no mussels but a heavy settlement of ascidians on all panels with the exception again of the galvanised weldmesh.

These results indicate that the exact conditions of emplacement and depth of immersion are critical for settlement and this must be considered when assessing the results of any trial settlement panels.

The settlement of ascidians and mussels on mesh netting, detailed above, imposes a severe additional increase in the weight of a mesh panel, especially if it is to be removed from the sea for cleaning and maintenance. From the above research an estimate of the increase in weight of surface nets month by month in the sea can be given (Milne, 1970e). It has been found that new nets inserted in the sea in the Northern hemisphere will gather slime films and weed up until June/July, but the settlement of marine organisms will greatly increase the weight by August/September as shown in Fig. 22 for seasonal weight increases.

SEASONAL WEIGHT INCREASES FOR FABRICS
(Milne, 1970e)

Fabric	Wt kg/m²	July	Multiplication Factor for Weight Increase	
			September	November
Nylon	0·23	2	85	108
Ulstron	0·34	2	64	110
Courlene	0·20	2	85	126
Polythene:				
(a) standard	0·18	2	112	200
(b) cupra-proofed	0·18	1	44	94
Netlon	0·34	1	36	48
Plastabond	3·25	0·75	10	13
Galvanised:				
(a) chain-link	2·03	0·3	0·5	2·75
(b) weldmesh	3·4	0·3	0·5	2·5

Fig. 22 Seasonal weight increases for fabrics due to marine fouling. (*Credit—From Milne, 1970e*)

These weights were measured in air after draining at each inspection of the panels. Fig. 22 shows that cupra-proofing a net halves the marine fouling,

and that the weight increase of galvanised fabrics is least of all. However, galvanised chain-link mesh cannot be recommended since abrasion at the links of the mesh tends to rub off the galvanising and start corrosion.

The other metallic materials have been used extensively in the past for marine use and knowledge of their fouling characteristics and corrosion has been investigated, for example, by Rogers (1968) who presents information in a useful manner and discusses the corrosion of metals in the sea. LaQue (1969) has also investigated the deterioration of metals in an ocean environment and his research is relevant to the use of metallic fabrics for mariculture. Many of the early applications of aluminium to marine use were not successful as unsuitable alloys were used. Recent research has shown that the magnesium alloys are more resistant to corrosion but, unfortunately, they initially attract large settlements of ascidians. Stainless steels, although very resistant to general corrosion, suffer attack from barnacles, which cause severe pitting of the metal. Brass, copper and nickel fabrics all suffer from the same problem that in sluggish or slow moving waters there is no resistance to the settlement of marine organisms, although the corrosion may be very slight.

Strength of Materials

To be able to select the best fabric for screening in marine farming a knowledge of the physical properties and elastic recovery under excessive loads is necessary. It is also essential to know the percentage strength retained in successive years to determine the useful life of the fabric.

Where netting stands above the water surface, either as a rigid structure to allow for tidal rise and fall, or as a floating net to allow for wave action and fish jumping, the climatic exposure of nets must be considered. Fortunately, several experiments have been carried out on the weathering of fibre materials and plastics to give an indication of the expected life and reduction in strength with time.

Little (1964) reports that the Shirley Institute carried out a study of the loss of strength in fibres of cotton, Terylene and nylon due to exposure to sunlight in the open air. After six hundred hours of sunlight, the percentage strengths of cotton, Terylene and nylon had fallen to 50%, 25% and 10% respectively.

A further experiment carried out at eight different sites in Britain by Little and Parsons (1967) showed that the reduction in fibre strength was related not only to sunlight, but also to climatic conditions.

Again cotton, Terylene and nylon were the fibres tested, and after one year the strength of the cotton was reduced to 20% in industrial areas and to 60% in clean unpolluted atmospheres. For Terylene and nylon the retained strengths ranged from 35–55% and 20–40% respectively.

More recent work has been carried out on man-made fibres as reported by Rosato and Schwartz (1968). They report that nylon is not recommended for outdoor exposure, as complete loss of strength would result in only six months if exposed to such excessive weathering as would be encountered in a marine environment. Pinner (1966) reports that polyethylene (such as Courlene) has a life of three years with a reduction in strength of 50%. Exhaustive marine research by Connolly (1963) of Bell Laboratories on 600 specimens, including natural and synthetic fibres, has also shown nylon to lose strength very quickly in the first year. Two out of four nylon specimens had pholad clam damage in the first year and half the polyvinyl chloride specimens also had pholad penetrations. The only specimens with satisfactory strengths were the vinyls with 50% after five years. One suggestion by Arzano (1959), who discusses the use of nylon and Courlene fish netting, is that to prevent loss of strength from the effects of sunlight and exposure the nets should be kept immersed at the site when not in use. However, both these types of netting are prone to severe marine fouling, as discussed earlier.

The elastic extension under load of the synthetic fibres (I.C.I. Fibres Ltd, 1968) is also important. The immediate percentage recovery from a 67% breaking load of Terylene, Ulstron and I.C.I. nylon is between 73–75%. However, for polyethylene (Courlene) the figure is only 41%. It is recommended that for design purposes a figure of 50% of the breaking load should be used in calculations to prevent too much extension under load (Milne, 1970e). The marine fouling research carried out by the University and S.M.B.A. described earlier on mesh netting panels has shown that chain-link mesh is not suitable for marine use as severe corrosion occurs at the links in the mesh. The vinyl cover on the Plastabond samples was worn through at the links in the first year, and the galvanised chain-link mesh was showing signs of deterioration at the links in the second year. Galvanised weldmesh, however, does not suffer from the above defect and will retain its strength for longer in the sea.

The use of galvanising to protect steel from corrosion is well known from ship research (Rogers, 1960) and so long as the weldmesh is hot dip gal-vanised to B.S. 729, Part 1, a life of three to five years (Butler and Ison, 1966) can be expected, depending on the exposure and water currents. It is essential that this quality of galvanising is obtained as inferior mesh will deteriorate rapidly, as found on one cheaper panel tested. If a material with a longer life expectancy is required the metallic fabrics, aluminium, stainless steel, brass, copper, cupro-nickel and nickel can be considered, as their physical properties make them very resistant to marine corrosion (Campbell, 1969; LaQue, 1969). However, owing to the unsuitability of chain-link mesh already mentioned, weldmesh panels would be necessary, and welding can sometimes prove troublesome with these metals.

If mesh netting is selected for the intertidal or sublittoral zone it will require to be anchored down to prevent chafing and abrasion owing to continual flapping caused by wind and wave action if not secured at regular intervals. Sample calculations of wind and wave loadings obtained from Appendices 1–3 are given in Appendix 4 for the design of net structures.

Mesh Maintenance and Cleaning

From the foregoing sections on marine fouling, strength of materials and fabric construction, it will be evident that various methods of cleaning the mesh will be required for different fabrics.

The synthetic fibre meshes, nylon, Terylene, Ulstron, Courlene, and polythene, on immersion in the sea initially attract growths of filamentous algae which can be cleaned easily, either underwater by divers or by removal with a hard-bristle hand brush. The cleaning of ascidians and mussels poses a more difficult problem owing to the massive settlements and increases in weight described earlier. If removal from the sea for cleaning is proposed, the net panels must be small and easily manageable to prevent damage on lifting out of the water. A brush is not suitable for the removal of ascidians from mesh panels, the alternatives being picking by hand or removal by a high velocity water jet. Once the main settlements are removed, the remaining byssus threads and barnacles can be removed by scrubbing with a hard-bristle hand brush. As the synthetic fibre meshes are flexible, care must be taken to avoid damage to the fibres.

The rigid-mesh nets, whether of polymeric or metallic fabric, are easier to clean, and a hard-bristle deck scrubber can be used both above and below the water line to remove both weed and mussels. Ascidians present a more difficult problem, as they require to be taken off by hand under water or by a high velocity water jet if the mesh is removed from the sea. If the underwater fouling is removed by divers, it is better that it should be done with a slight current flowing, rather than at high or low water, when the debris in suspension can reduce the diving visibility to nil.

The removal of floating drift weed, bladder wrack and *Laminaria*, can be accomplished at low water by boat using a wooden or plastic rake on rigid meshes, but only by hand on flexible synthetic-fibre meshes, because of the danger of tearing the mesh if a rake is used.

It is essential that weed and marine organisms, be they ascidians or mussels, be completely removed from the area of the mesh netting, and deposited where they are unlikely to be brought back by the tide.

The selection of the right mesh and its antifouling properties can be critical for the maintenance of the enclosure, and establishment of a commercial concern. For example on one yellow-tail fish farm at Megishima in Japan, out of a total staff of 25 on the farm, 7 were divers employed full-time on the cleaning of the 1,370 m length of netting, Fig. 14. At another yellow-tail fish farm at Matsumigauru, also in Japan, the mesh netting was so fouled with ascidians that the net was carried away by the water flow during a storm, with complete loss of stock.

Many of the early net enclosure fish farms in Japan (Allen 1968), used a variety of mesh fabrics, synthetic fibre, vinyl covered wire and galvanised wire. At a recent (1971) meeting of the author with several Japanese fish culturists, it was discovered that as a result of the original trial and error developments in Japan, using various types of mesh, galvanised mesh was now recognized as the best material due to its anti-fouling properties, with the need for less maintenance and thus fewer divers. This concurred with the marine fouling experiments carried out in Scotland (Milne and Powell, 1972), before the construction of large fish farms with expensive mistakes.

Predator and Trash Net

The selection of the best type of mesh fabric for the retention of fish in a marine environment has been discussed earlier from the viewpoint of length of service and resistance to marine fouling. However, damage to the net from outside predators or floating trash such as timber planks or tree branches must also be considered.

In the case of a light fibre retention fish net, where damage could cause the loss of all the fish,

it is recommended that the nets should be protected by an outer net of a larger mesh. This predator and trash net would protect against predators such as seals and also deflect floating timber and other trash.

The installation of two nets on one net structure can pose several problems with regard to maintenance and operation, owing to the abrasion and snagging of the nets if they are hung in contact with one another. The nets must, therefore, be hung sufficiently far apart to prevent any abrasion and snagging. The only known double net floating cages are used by the Japanese in exposed situations for the cultivation of yellow-tail as described in Chapter 10. With piled structures, the two nets can be hung on either side of the framework, but with a moored system the nets must be hung on separate lines of floats, or suspended from either side of a pontoon. Mooring systems are not generally suitable for shallow water owing to the influence of tidal currents causing the floats to drift with the flood and ebb tides.

If a heavier fibre or rigid fish retention net is used, a trash net will be required only in exposed situations where the combination of the trash with wave action could cause damage. The type of fish netting selected and whether or not a predator and trash net is required will determine the best method of installation for the netting to form the marine enclosure. The choice of method will depend on the design conditions pertaining to the site selected and the operational and maintenance loadings.

At the Faery Isles fish enclosures in Loch Sween, discussed in Chapter 8, where 25 mm galvanised weldmesh is used as the retention net, a predator and trash net consisting of 75 mm galvanised weldmesh is provided only on the offshore face of the enclosure, Fig. 23.

Surface Floating Units

One of the major aspects in the expansion of sea farming in the last decade has been the potential increase in available sea space for farming due to the development of surface floating techniques.

The early rafts used old steel oil drums for flotation, and a timber framework, often of bamboo. However, these rafts required continuous maintenance with the replacement of drums and timber. The production of large fibreglass drums, polystyrene shapes, and styrofoam cylinders has transformed the methods of flotation and reduced the replacement costs. For example, the rafts used for the marine fouling studies described earlier in

Fig. 23   Predator and trash net of 75 mm galvanised weldmesh fixed on offshore face of Faery Isles Fish Enclosures, Loch Sween. 25 mm galvanised weldmesh is used for the retention of the fish.   (*Credit—P. H. Milne*)

Fig. 24   Use of polydrums for raft flotation. The lightweight raft framework was constructed from 50 mm alloy scaffolding.   (*Credit—P. H. Milne*)

this chapter used polydrums for flotation, with a framework of 50 mm scaffolding, Fig. 24. However, sometimes the fouling of these containers can be severe, and on one Japanese farm this fouling has been combated by ensheathing the drums in polythene bags which are thrown away each year, thus reducing the cleaning and maintenance costs.

The development of lightweight ferro-cement containers also heralds a new technique in providing a flotation unit with anti-fouling properties.

The selection of the material for the structure of the raft framework should also be considered since some waters have prolific wood borers, which could result in total deterioration of the raft. One such example occurred with rafts constructed from bamboo in Venezuela for the cultivation of mussels where 122 rafts were written off as a total loss.

Timber can, however, be protected against the ravages of wood borers and *Teredo*, and in Australia the oyster farmers dip their spat settling sticks in tar before immersion in the sea for this reason.

3

SEA FARM CONSTRUCTION TECHNIQUES

Chapter 6
Design and Construction of Shore Facilities

The design and construction of hatcheries, ponds, raceways and tanks for sea farming on the shore enables some environmental control over the species to be farmed. The construction of such shore facilities is very similar to those existing for fresh water fish farming throughout the world. However, in selecting a shore site for construction, the associated hydraulic and pumping requirements should be fully considered.

The source of sea water for hatcheries, etc., is most important since pollution of various kinds, as discussed in Chapter 13, can seriously affect the survival and growth of larvae. Davis (1969) describes the development of shellfish hatcheries and says that the water quality should be scrupulously checked in areas of heavy domestic or industrial pollution, or heavy run-off from intensively pesticide-treated areas, and areas where intensive, possibly toxic natural blooms occur regularly. Davis concludes that the future direction or directions which hatchery production of shellfish may take are limited only by the imagination of the industry. However, he warns that with increasing urbanisation, additional areas will be rendered unsuitable for the larval stages to develop, and hatchery rearing through the critical larval stages will become increasingly important in many places. The same can be said to apply to fish, and accordingly many laboratories throughout the world are working on artificial cultivation and spawning techniques for the production of fry, as described in this chapter.

Fig. 25   In the foreground 4·5 m dia. salmon ponds 1 m deep at Otter Ferry, Loch Fyne. In the background small smolt production tanks for salmon. Fresh water supply channel in middle. (*Credit—P. H. Milne*)

Fig. 26 Concrete raceways for fish farming. This type of raceway is suitable for either rainbow trout or salmon culture.
(Credit—by courtesy of A/S Mowi)

Several types of shore ponds are mentioned in the text for fish farming. Circular ponds are generally shallow, 0·6 m–1 m, compared with their diameter, 5–10 m, with the inflow at a point round the periphery, and the outlet in the centre, Fig. 25. Raceways are long, narrow ponds, with average dimensions, 2·5 × 30 × 1·0 m, where the water inlet and outlet are at opposite ends, Fig. 26 (Buss *et al.*, 1971). The Foster-Lucas pond is a hybrid of circular and raceway ponds, having straight sides and circular ends with average dimensions, 5 × 23 × 1 m. These ponds have a centre baffle wall with the inlet and outlet on either side. In selecting a pond design for fish farming it is most important to realise that the biological and physical conditions in the ponds are dependent on the inherent hydraulic characteristics (Burrows *et al.*, 1955).

(I) MOLLUSCS

Oyster Culture on Long Island, United States of America

The culture of the American oyster, *Crassostrea virginica* (Galtsoff, 1964), varies widely throughout the United States, but in recent years there has been a decline of the industry due to problems of disease and pollution changing the ecology of the oysters' habitat. This summary of oyster culture at Long Island, New York, describes some of the modern techniques now being used to combat these problems. Of the four commercial oyster hatcheries at Long Island, only that of G. Vanderborgh and Sons will be discussed in detail, but all operate under similar conditions (Ryther and Bardach, 1968). One company using an open water spawning technique is described in Chapter 8.

Sexually mature adult oysters are selected from the growing beds and are maintained at 10°C in the hatchery. They are then conditioned for spawning by raising the temperature slowly to 18°C. After 2–4 weeks spawning is induced by further raising the temperature to 25°C. The fertilised eggs obtained are then transferred to a battery of fifty conical 450 litre rearing tanks. The water is drained through the bottom of these tanks every other day and the larvae caught in a fine mesh screen. These larvae are then graded and only the larger, more rapidly growing individuals, approximately 20% of the total, are kept for cultivation. These selected larvae are then transferred to larval rearing tanks and are fed with algal cultures. After 10–15 days the larvae are ready for setting and are again transferred to six 3,000 litre plastic settling tanks which have a layer of thoroughly cleaned oyster shells on the bottom, to which the larvae can attach as they metamorphose (Davis, 1969).

Once the oysters have set, which takes from 24–48 hours, the shells with the attached spat are transferred to $\frac{1}{2}$ bushel (1 bu = 36·4 l) plastic mesh bags, which are suspended from wooden beams into nursery tanks. Each of the 22,500 litre cement nursery tanks is capable of holding 200 of these $\frac{1}{2}$ bushel bags, and is situated in a greenhouse type habitat with the temperature kept at 30°C. The feeding of the oyster spat with algal cultures is continued here, and with the high temperature, a very rapid growth of the young seed oysters is obtained. The oysters are kept in these nursery holding tanks from 4–7 days before transfer to the open sea. The

Fig. 27   Transfer of shell bags with oyster spat by chain hoist from the Vanderborgh hatchery to floating rafts at the dockside. (*Credit—Bureau of Commercial Fisheries*)

Vanderborgh hatchery is built on the dockside at Long Island, so the wooden beams with shell-bags attached can be lifted by chain hoist and overhead rail outside to the dock, Fig. 27, where they are placed on a floating raft, each raft also holding 200 bags, for towing to a bay site. The seed oysters are only kept on the rafts for 2–3 weeks until they reach 1–2 cm when they are ready for transplanting

to the oyster beds, as detailed in Chapter 9. Due to cold temperatures below 10°C in the winter when the oyster seed cannot be planted, this hatchery operation is restricted to late spring and summer.

Oyster Culture in Britain

The native British oyster is *Ostrea edulis*, which lives and spawns round the coastline; more predominantly around the English coasts where the summer water temperature rises above 15°C. However, due to disease, pollution, dredging and hard weather, these stocks have been devastated. For example, at Billingsgate alone, it is reported that some five hundred million oysters were sold in 1864. Today, a century later, not more than eight million are produced throughout the whole of Britain (Yonge, 1966b, 1970).

To replace these oyster beds at the beginning of the 20th century oyster spat of the Portuguese oyster, *C. angulata*, from France, and also of the American oyster, *Crassostrea virginica*, were imported and relaid on the English oyster beds. The import of European stocks helped to improve the oyster fishery, but the import of the American oysters introduced oyster predators, unknown in Britain, which wiped out many indigenous oyster beds. Another problem has been the variability of oyster spat for collection in the coastal waters; hence the need for the importation of oyster spat for relaying.

Since the second world war, both the Ministry of Agriculture, Fisheries and Food and the White Fish Authority have been carrying out research in an attempt to improve the prospects of oyster cultivation in Britain. The first stage in modern cultivation methods consists of inducing the oysters to spawn under controlled laboratory conditions regardless of natural temperature changes, which restrict spawning to the summer months (Loosanoff and Davis, 1963).

Seed oyster technology in Britain has been primarily developed at the Conway Laboratory of the Ministry of Agriculture, Fisheries and Food, into the production of reliable methods for the culture of larvae of *O. edulis* (Walne, 1956, 1965, 1966, 1969, 1970a,b, 1972). These techniques have been further developed and extended by the White Fish Authority with a hatchery also at Conway, into an effective costed, commercial proposition (Knowles, 1968, 1972; Richardson, 1969; Anon., 1972).

As oysters are filter feeders, they require a supply of unicellular algae for food. The White Fish Authority's unit at Conway has now developed and patented a system for the high-density cultivation of these algae especially for oyster hatchery production (Knowles *et al.*, 1971). At Conway the spawning of the *O. edulis* breeding stock is induced by raising the water temperature in shallow trays to about 20°C. After the two to three week free-swimming period the larvae settle and are thereafter termed spat which can be used for relaying on oyster beds or for tray culture.

To determine whether oysters could be produced on a commercial basis employing the above methods, the White Fish Authority has devised a linear programming model, which has been used to find the most profitable method of utilising production facilities at the present state of development (Haywood *et al.*, 1970).

In addition to the intensive development of hatchery production techniques in Britain, studies by the Ministry of Agriculture, Fisheries and Food are also being carried out into the density of laying oysters in trays, tray design, and growth and survival on and above the bottom (Key, 1970). The trays used for these experiments were 1 m square and 5 cm deep, covered in 1 cm weldmesh. It was found that oysters on trays raised 0·5 m on stakes had a lower mortality than those grown on the river bed.

To assist commercial oyster farmers in the cultivation of the Pacific oyster, *Crassostrea gigas*, work commenced in the Conway Laboratory of the Ministry of Agriculture, Fisheries and Food in 1965, with oyster spat imported from British Columbia. Experiments both at the hatchery and in the sea have shown that the Pacific oyster can be grown to marketable size in British waters in 2–3 years (Walne *et al.*, 1971).

As a result of this research work, commercial concerns are establishing hatcheries for the production of oyster seed using these methods. Two hatcheries have now been constructed, the first at Loch Creran, Argyll in Scotland, and the second at Poole, Dorset in England.

The first new commercial oyster hatchery, owned by Scottish Sea Farms Ltd at Loch Creran, a sea loch just north of Oban, Fig. 18, was completed in January 1970 (Milne, 1970c). The process begins with the parent oyster spawning under controlled environmental conditions in the hatchery. The free-swimming larvae are kept for two weeks in specially designed vats and fed on cultivated plankton. When the oysters grow to 5–10 mm in size they are ready for transplanting to prepared sea beds. After about six to nine months, depending on the season, the oyster seeds are ready for the seed market and mature oysters are available for market-

ing three to four years later. Maximum use is made at Loch Creran of the latest scientific methods, and skin divers are employed when the young oysters are transplanted on prepared sea beds.

The hatchery, when in full operation, is aimed at the high potential export market to France, Holland, Belgium and Italy. The first consignment of seeds prepared in 1970 were despatched in the Spring of 1971; the initial target of this Company is four million seed oysters per annum.

The second commercial concern to establish large-scale hatchery techniques for oyster seed production has been Poole Oyster Co. at Poole in Dorset. This company are particularly concerned with the development of an adequate and regular supply of oyster seed. With its specialised knowledge of genetics, the company hopes to improve the breeding and production methods of oysters. The planned capacity of the hatchery is 12 million seed oysters per annum; four million of which will be kept for stocking their own oyster beds, and the remainder will be sold as seed (Anon., 1970f). It is also proposed to investigate and improve methods of mechanical harvesting using especially equipped dredgers.

Abalone Culture in Japan

Due to the heavy demand for the northern Japanese abalone, *Haliotis discus*, cultivation attempts are being made in Japan. At present the young juvenile hatchery abalone, between 15–20 mm in size, are sold directly to the Fisherman's Cooperative Association, for sowing on the seabed for harvesting after about three years when they reach marketable size (Ryther and Bardach, 1968).

The original research into the cultivation of the abalone was initiated by Dr T. Imai, the Director of the Oyster Research Institute at Kesennuma. The production of the abalone seed for stocking is, however, carried out at the Kanagawa Prefectural Fisheries Experimental Station. Here sexually mature specimens are placed in an outside concrete tank, $2 \times 2 \times 0·5$ m for spawning. The fertilised eggs produced are then transferred to deeper concrete tanks, $2 \times 1·2 \times 1·2$ m inside the laboratory at a density of 100,000 eggs per tank for the 8–11 day larval period. Just before the abalone metamorphose and settle, collectors of corrugated plastic sheets, $0·5 \times 0·5$ m are placed vertically into the tanks. Prior to insertion in the settling tank the plastic sheets are carefully maintained in running sea water in order to develop a surface coating of benthic diatoms. Once the abalone have settled on these collectors they are moved out-

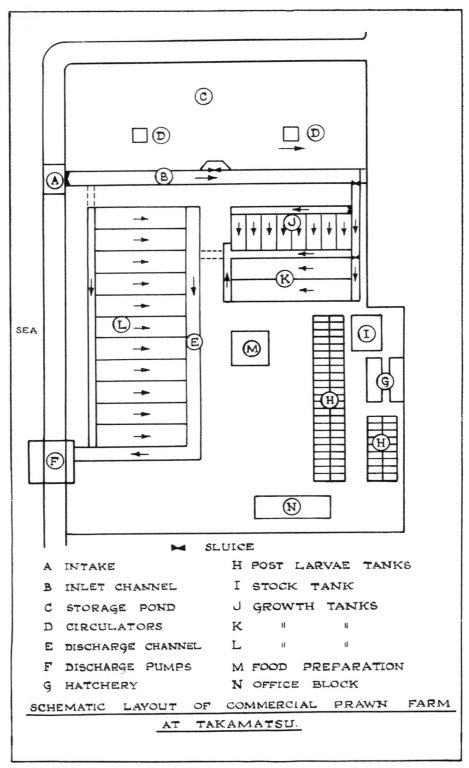

SLUICE

A INTAKE	H POST LARVAE TANKS
B INLET CHANNEL	I STOCK TANK
C STORAGE POND	J GROWTH TANKS
D CIRCULATORS	K " "
E DISCHARGE CHANNEL	L " "
F DISCHARGE PUMPS	M FOOD PREPARATION
G HATCHERY	N OFFICE BLOCK

SCHEMATIC LAYOUT OF COMMERCIAL PRAWN FARM AT TAKAMATSU.

Fig. 28 Plan of the shrimp farm owned by the Kuruma Shrimp Farming Co. Ltd at Takamatsu in Japan.

(*Credit—From Allen, 1968*)

doors again and submerged in tanks, 9 × 9 × 1·5 m tanks with running sea water. One thousand collectors, each with approximately 10,000 abalone are placed in each tank. Once the young abalone have reached 5–10 mm they are thinned to 80 per sheet for maximum growth. After eight months the juvenile abalone have grown to 15–20 mm, and are then ready for stocking.

There are no technical or biological problems to prevent the pond culture of abalone to mature market size. At present it **is** only the space involved during the three years to harvesting that is impractical. Economic techniques for the holding of abalones and the possible use of warm water from coastal power stations could change this situation considerably and increase its commercial prospects.

(II) CRUSTACEANS

Shrimp Culture in Japan

The commercial cultivation of the Kuruma shrimp, *Penaeus japonicus*, can be traced back to the pains-

taking research of one man, Dr Motosaku Fujinaga (Hudinaga). He started work on the artificial spawning and hatching of the Kuruma shrimp in the laboratory in 1933 (Hudinaga, 1942). Initially difficulties were experienced in the feeding of the shrimp in the zoeal stages. By 1941 it was found that the diatom, *Skeletonema costatum*, provided suitable food for the zoeal shrimp, and work continued. Unfortunately Dr Fujinaga's laboratory at Aio was destroyed by a typhoon in 1942 and much of the experimental data and results, etc., was lost. It was not until after the end of the second world war that he could continue his research into the cultivation of the Kuruma shrimp. However, the feeding of the zoeal shrimp through mysis up to the early post-larval stage proved a problem until it was discovered that the nauplii of brine shrimp, *Artemia*, could be used as food. Once this hurdle had been overcome the commercial exploitation of *P. japonicus* was possible and Dr Fujinaga resigned his governmental post to establish his own company, the Kuruma Shrimp Farming Co. Ltd at Takamatsu

Fig. 29   Large concrete outdoor tanks used by the Kuruma Shrimp Farming Co. Ltd at Takamatsu in Japan.

(Credit—J. H. Allen)

in 1959 (Hudinaga and Miyamura, 1962), Fig. 28. The scale of operations has greatly increased in the last decade and the cost of production lowered by the construction of large outdoor tanks filled with natural sea water, Fig. 29 (Fujinaga, 1969). There has been such a rapid development in the commercial technology of cultivating the Kuruma shrimp that it might be useful to discuss the various stages of development in the last ten years, up to the current technique.

At first 80 small wooden tanks, 2 m × 1 m × 1 m, were used for the spawning, hatching and raising of Kuruma shrimp larvae. These tanks were kept indoors with little light and supplied with filtered sea water. The larvae were first fed with algal cultures like *S. costatum* at the zoeal stage, and then with brine shrimps at the mysis stage. After the post-larval stages the food was changed to finely crushed meat of short-necked clams.

However, the above system was very expensive, and the method was improved in 1964 to reduce costs, and concrete tanks, 10 m × 10 m × 2 m, each holding an average 1,500,000 fry, were successfully used. Some of the tanks did not receive enough aeration with a subsequent drop in production. This is one of the problems of shrimp culture in the provision of adequate aeration at the tank bottom. To overcome this problem, 18 special tanks, 50 m × 10 m × 1·3 m, were constructed at Takamatsu with double bottoms. The false bottom consists of fine plastic mesh supported on split bamboo canes fastened 10 cm above the bottom of the tank. The mesh is then covered with 75 mm of clean sand and an air-lift system circulates the sea water through the sand (Hudinaga and Kittaka, 1966). The shrimps from the hatchery tanks were transferred first to post-larval concrete holding tanks approximately 5 m square, and second to the double bottomed tanks. This intermediate step was cut out in 1966 and the post-larval tanks rendered obsolete. In addition to the above, 28 tanks measuring 100 m × 10 m × 0·6 m with sand bottoms were also used, water being pumped from the sea to give

Fig. 30 Large indoor concrete tanks 2 × 6 × 1 m for rearing the shrimps at Takamatsu up to 2 cm.
(Credit—by courtesy of the White Fish Authority)

Fig. 31 Small concrete outdoor tanks now rendered obsolete with recent developments by the Kuruma Shrimp Farming Co. Ltd, at Takamatsu in Japan. (*Credit—J. D. M. Gordon*)

a continuous flow (Ryther and Bardach, 1968).

By 1970 the commercial production of *P. japonicus* had become streamlined and simplified. The zoeal stage now use the original 2 m × 1 m × 1 m indoor tanks and are fed on algal cultures. They are then moved to larger indoor tanks 2 m × 6 m × 1 m, Fig. 30, and reared to the 2 cm stage on brine shrimps and later minced clam meats. On reaching the 2 cm size they are transferred straight to a large intertidal pond, 4 ha in area and 1 metre deep, as detailed in Chapter 7, Fig. 31 (Williamson, 1971). The other concrete and false bottom ponds at Takamatsu would now appear to be obsolete. Takamatsu is not, however, the only shrimp farm to be developed, as other commercial companies are now involved in similar activities following Dr Fujinaga's lead. The history of Takamatsu has been discussed in detail to show how techniques can change in such a relatively short time.

In addition several governmental organisations, federal, regional and prefectural are also engaged in rearing juvenile shrimps in hatcheries for selling to farm organisations or for release into the sea. One example is the Yamaguchi Prefectural Station situated behind a coastal embankment at the western end of the Inland Sea, which was established in 1963 with top priority for shrimp cultivation. The major facility is the hatchery with 178 tanks, each 3 m × 1½ m × 1 m. In addition there are also 384 concrete tanks of various sizes up to 6 m × 18 m × 0·6 m for larvae rearing, covering an area of 1·2 ha. To allow for colder winter temperatures, 19 tanks, 1 m deep and up to 60 m × 20 m in size, are provided for over-wintering, covering an area of 2·4 ha. The sea water supply is pumped into a reservoir (2·4 ha) at high water, and is fed to all the tanks by gravity. Drainage channels then carry the water back to a discharge pond where it is pumped out.

The Japanese willingness to invest capital in shrimp cultivation is obvious at establishments such as Takamatsu and Yamaguchi. However, Ryther and Bardach (1968) having examined the economics of such facilities have concluded that this method of shrimp culture could only be economical in Japan. The relatively low cost of labour and raw materials balanced against an extremely high value for the product, are cited as the criteria for profitability. In fact Dr Fujinaga's (1969) current research is largely directed towards finding suitable foods, other than potential human foods, for juvenile shrimps, and this could increase its adaption elsewhere.

Shrimp Culture in the United States of America

The Japanese methods of shrimp larvae culture developed by Dr Fujinaga, detailed earlier in this chapter, have now been transposed to several laboratories on the Gulf Coast of the United States. These techniques have generally been successful with the Gulf of Mexico species: the pink shrimp, *Penaeus duorarum*; the brown shrimp, *P. aztecus*; and the white shrimp, *P. setiferus* (Webber, 1970).

The cultivation of shrimp larvae is at present being carried on at several government, university, and industrial laboratories on the Gulf coast. The large-scale cultivation of larvae is most actively being investigated by Cook (1969) at the Bureau of Commercial Fisheries Laboratory in Galveston, Texas; Idyll *et al.* (1969), at the Institute of Marine Sciences, University of Miami, Florida; and Miyamura of Marifarms Inc., Florida (Anon., 1970c).

Cook commenced his work on raising shrimp larvae using the techniques of Fujinaga (Hudinaga, 1942), but found he could not rear penaeids in the United States to the post-larval stage. Initial research was carried out with the pink and brown shrimps, as these were readily available for spawning, and after modifications, some success was experienced (Cook, 1966; Cook and Murphy, 1966). This work has now continued to the stage that further developments to the early research have enabled large numbers of brown and pink shrimps to be reared for pond stocking (Cook, 1969; Cook and Murphy, 1969, 1971). Only small numbers of the white shrimp have so far been reared due to the difficulty experienced in obtaining gravid females for large-scale spawning techniques.

The supply of larvae is still dependent on the commercial trawling to capture female shrimps in spawning conditions. These shrimps are then transported from the vessel in 75 litre plastic barrels to the laboratory, where they are placed in 946 litre fibreglass tanks (Marvin, 1964) containing

560 litres of sea water. Three shrimps are allocated to each tank and after the eggs have hatched the nauplii are siphoned off and placed in an inverted 19 litre polyurethane carboy with the bottom removed. These carboys contain 15 litres of filtered, aerated sea water, half of which is replaced daily, and can be used to rear up to 4,000 larvae to the post-larval stage. As in the Japanese hatchery techniques, the diatom, *Skeletonema costatum*, is used as food at the zoeal stage and newly hatched brine shrimps, *Artemia* sp., at the mysis stage. The range of temperature and salinity is also critical for larvae survival and Cook (1969) recommends temperatures in the range 28–30°C and salinities between 27–35‰.

The University of Miami has also been interested in the cultivation of shrimps (Idyll, 1965) and started work in 1967 on long-term research into the intensive culture of the pink shrimp, *P. duorarum* (Idyll *et al.*, 1969). A laboratory has been built in South Dade County, Florida on the grounds of an electrical generating plant owned by the Florida Power and Light Co., for research into the larval culture and propagation of larval food organisms. This work has been sponsored under the Sea Grant Programme. Initial experimentation has involved the use of the wild pink shrimp as a source of eggs, but a selective breeding programme is under way. Adjacent to the hatchery are 16 outdoor concrete larval rearing tanks each having a capacity of 20 tonnes of sea water. The 2 hectare site also contains one 0·4 ha, two 0·2 ha, and four 0·1 ha ponds (Tabb *et al.*, 1969) for shrimp studies.

These ponds were constructed between two man-

Fig. 32 Shrimp ponds in Florida. The pond levees are graded steeply and lined with Uniroyal EPDM membrane. (*Credit—D. C. Tabb, U.S. Dept of Commerce, Sea Grant Program*)

Fig. 33 Top edge of pond liner fixed to timber batten 5 × 20 cm, to be buried in ditch at crown of levee for shrimp ponds in Florida. (*Credit—D. C. Tabb, U.S. Dept of Commerce, Sea Grant Program*)

made canals and water can be obtained from both of these to give either ambient sea water temperatures or power station outfall temperatures. The ponds are all built with their bottoms 0·6 m above mean high water, thus allowing drainage at all states of the tide. However, it means that two pumps rated at 0·04 m³/s are required to ensure adequate water movement and circulation. These two pumps are capable of filling the 0·4 ha pond to a depth of 1·5 m in 24 hours. This pump capacity also makes it possible to supply sea water to each pond to offset losses due to evaporation and seepage and to supply oxygenation with warming or cooling as required.

Since the ponds were constructed from crushed limestone it was necessary to line the ponds to make them watertight. This was effected using a Uniroyal EPDM membrane, 2·5 mm thick, to line the ponds as shown in Fig. 32. The pond levees were graded steeply to discourage the perching of wading birds since they can be serious predators, especially with low water levels. This liner was only used on the levee slopes. The bottom edge was sealed by running it 0·6 m under the pond bottom, which consists of 10–15 cm layers of calcium carbonate marl. The top edge of the liner was fixed to timber battens 5 × 20 cm sunk in ditches along the tops of the levees, Fig. 33. An outlet drain stack is shown in Fig. 34 with the pond filled ready for stocking. These liners were installed when the ponds were constructed in February 1968 and have served very well since then. These ponds are also used for pompano as discussed later in this chapter.

Marifarms Inc. (formerly called Akima International Inc.) are, however, a new development on the scene in the United States. Marifarms was established in 1967 and started an extensive research and development programme in the Spring of 1968 in Panama City, Florida, under the direction of Dr M. Miyamura, formerly with the Kuruma Shrimp Farming Co. Ltd in Japan. The Japanese techniques and expertise of obtaining shrimp eggs and bringing the shrimps to a mature size in an economical time have been developed in Panama City. They succeeded in hatching 10 million shrimps in 1968 and 20 million in 1969, with 400 million in 1971 (Anon., 1971n).

Experiments have also shown that mature shrimps can be raised in six to eight months, and the company harvested some 226 tonnes of shrimp in 1971 from their ponds and enclosures. Of the 400 million shrimps hatched in 1971, the company released 378 million to their own enclosures and 22 million to the open waters of West Bay. This public release is a legal requirement of Marifarms' lease of the sea for its 1,000 hectare enclosure in West Bay, Florida, as discussed in Chapter 8.

Shrimp Culture in England

A number of cold water species of shrimp occur in European waters, and one of these is the English prawn, *Palaemon serratus*. Early research on this species by the Ministry of Agriculture, Fisheries

Fig. 34 Outlet drain stack in shrimp ponds in Florida with the pond at its operational level. (*Credit—D. C. Tabb, U.S. Dept of Commerce, Sea Grant Program*)

and Food, at their Fishery Experiment Station in North Wales, demonstrated methods of obtaining larvae from mature adults in the laboratory (Reeve, 1969a,b,c; Forster, 1970). The ecology and environmental conditions were also studied and although not considered for cold water cultivation, *P. serratus* was thought to be suitable for warm water cultivation.

The first experimental shrimp farm to be constructed in Britain has therefore been built alongside a coastal power station at Hinkley Point in Somerset. The farm consists of a hatchery building and three growing ponds, 15 m × 5 m in area, Fig. 35. Butyl rubber sheeting was used for the pond liners as shown in Fig. 36. Water is pumped from the power station outlet channel to a circular settling tank adjacent to the ponds, Figs. 35–36, to remove any particulate matter in suspension, especially silt.

In 1971 two populations of English prawns were grown from the post-larval to marketable size and various ways of feeding were tested. These experiments proved very successful and the next step is to achieve raising the prawns consistently to marketable size on a larger scale.

Lobster Culture in the United States of America

Hatcheries for the rearing of larvae from the American lobster, *Homarus americanus*, have been in existence since the beginning of the nineteenth century (Iversen, 1968). Although the biological knowledge necessary for raising these lobsters from the egg to the adult, which takes a minimum of 5–6 years in nature, has been known for some time, the farming of lobsters has been impractical due to this long growth period.

However, recent research by the Bureau of Commercial Fisheries, at their Biological Laboratory at West Boothbay Harbour, Maine, may well lead the way to modern techniques for farming the lobster (Kensler, 1970). This research has investigated the effect of increased temperature on the metabolism of the lobster throughout its life cycle. It was found that the eggs of egg-carrying females maintained in temperatures up to 26°C, hatched in half the normal 10 months it takes in nature.

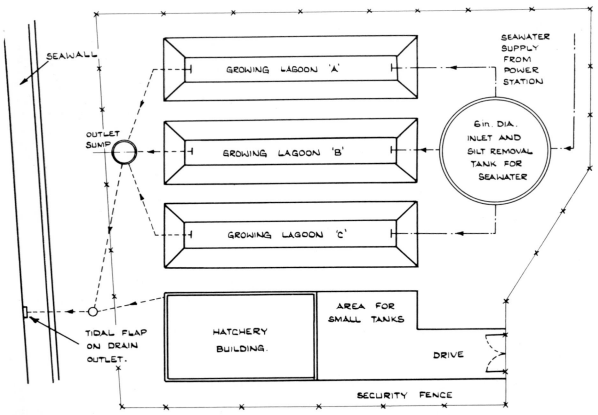

Fig. 35 Layout of experimental shrimp farm constructed adjacent to Hinkley Point Power Station in Britain.

(Credit—by courtesy of M. Ingram)

Fig. 36 Shrimp ponds at Hinkley Point using butyl rubber pond liners. The power station and circular silt settling tank are in the background. (*Credit—Central Electricity Generating Board, by courtesy of M. Ingram*)

Increased temperatures also accelerated the moulting frequency and thus promoted faster growth of the lobsters. The survival of the larvae was also much better at these higher temperatures.

In tests carried out at the Massachusetts Lobster Hatchery it was possible under controlled temperature conditions of 22°C with daily feeding to raise a one-year-old lobster (carapace length: 25 mm) to a marketable weight of 0·4 kg (carapace length: 75 mm) in 15 months (Shlesser, 1971). These tests indicate that with feeding and controlled temperature conditions, the time to raise a 0·4 kg lobster can be reduced from 5–6 years to about two years. In addition, by a selective breeding and mating programme, using warm water, it has been possible to hatch eggs every day of the year. This research has paved the way for the commercial farming of lob-

sters from the egg to marketable size in captivity, with the possibility of utilising the warm water effluent from coastal power plants.

(III) FISH

Rainbow Trout Culture in Norway

Although the American rainbow trout, *Salmo gairdnerii*, hatches in fresh water it may be acclimatised to sea water for growing and fattening in commercial cultivation.

The Vik brothers (Vik, 1963) carried out early research into this cultivation acclimatisation and had three pools constructed; one fresh, one salt and an intermediate "accustoming" pool for transfer between the fresh and salt water pools. They discovered that by accident some rainbow trout had escaped from the fresh water pool and had made their way through the accustoming pool into the sea water pool where they had grown rapidly to a marketable size. After a series of experiments to determine the rate of change from fresh to salt water it was possible to consider the commercial production of rainbow trout in sea water.

The original reason for the use of sheltered salt water Norwegian fiords, rather than fresh water, as in Denmark (Sedgwick, 1966), was lack of suitable streams and the prevailing cold water temperatures. Sedgwick (1970) has summarised the advantages of using salt water for rainbow trout culture:

(i) Sea water areas influenced by the Gulf Stream have smaller temperature fluctua-tions than fresh water and consequently the fish will feed for a longer period of the year and thus grow more quickly;

(ii) There is less disease risk to rainbow trout in sea water;

(iii) The fish can be kept in sea water at a density up to approximately ten times that of fresh water, thus saving on capital expenditure for retention, provided the water flow is also increased ten times; and

(iv) It is said that the fish make a better conver-sion of supplied food in sea water compared to fresh water.

The location of many of the Norwegian farms has been decided on the proximity to a fishing port

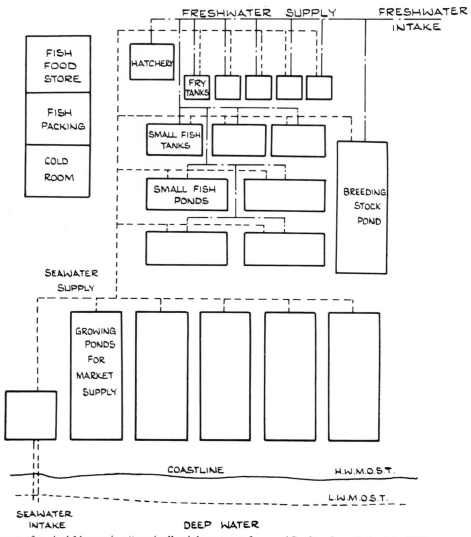

Fig. 37 Layout of typical Norwegian "marine" rainbow trout farm. (*Credit—from Sedgwick, 1970*)

where a constant supply of cheap sea food is available for feeding to the rainbow trout. In general these farms fall into two categories; first, those constructed on the shore with sea water pumped through the ponds or raceways, and second, those using enclosures in the sea, as dealt with in Chapters 8 and 10. Out of 70 rainbow trout farms developed in Norway in the last decade, 21 used sea water ponds on land and 35 were enclosures in the sea (Berge, 1968).

The typical layout of a Norwegian "marine" rainbow trout farm is shown on Fig. 37 (after Sedgwick, 1970). Here each of the ponds can be supplied with both fresh and salt water to allow the salinity to be varied, as the rainbow trout are acclimatised from fresh to salt water. Twenty-one farms of this type are in existence; some constructed with earthen ponds and others with concrete ponds. Earth ponds of course are less expensive to build than concrete ponds, but suffer from the disadvantage that they are more difficult to keep

clean and disinfect. The earth pond is also more liable to deteriorate with consequently more expenditure on maintenance. Some farmers allow the water to flow through several ponds before discharge, but this is not advisable due to the risk of spreading disease, and this risk can be avoided if the plant is correctly designed. These 21 farms cover an area of approximately 55 hectares. The annual production total from rainbow trout farming in Norway in 1971 was estimated at 1,000 tonnes.

Rainbow Trout Culture in Scotland

Following the success of the Vik brothers' experiments in rearing rainbow trout, *Salmo gairdnerii*, in sea water in Norway (Vik, 1963), three "marine" rainbow trout farms have been established in Scotland.

Marine Harvest Limited, a subsidiary of Unilever, built the first one at Lochailort, Inverness-shire in 1966 (Scott, 1968). The site chosen lay on flat ground just above the high water spring tide mark

Fig. 38   Concrete holding ponds for rainbow trout and salmon at Lochailort, Inverness-shire in Scotland.

(Credit—P. H. Milne)

Fig. 39 Small holding tank for rainbow trout. The salinity of the water supply can be varied as the fish grow for acclimatisation to salt water. (*Credit—P. H. Milne*)

adjacent to the River Ailort. Fourteen holding ponds for the fish were constructed in concrete with a central catwalk for access and water supply, Fig. 38 (Milne, 1970c, 1971a). By damming the River Ailort, a fresh water holding pond was created and fresh water pumped to the site. A pump-house was later established on the shore to supply the farm with a steady supply of sea water. After hatching, the young are transferred to the concrete holding tanks, Fig. 39, in early summer, and gradually acclimatised from fresh water to full sea water (Sedgwick, 1970). The salinity of this water can be closely controlled until the fish are in full sea water. Once this stage has been reached the fish no longer need to be kept in an expensive controlled tank environment on the shore, and the fish are transferred to sea cages floating offshore in Loch Ailort, as detailed later in Chapter 10.

Howietown and Northern Fisheries Limited were the second organisation to consider farming rainbow trout in sea water. However, they wished to transfer the fry straight from the hatchery into the sea, and thus chose an area with salinities varying from 20–30‰. In addition they wished to utilise the tidal range to stimulate circulation but without a floating facility. Therefore in 1968 they chose a sublittoral enclosure technique similar to

that being studied at that time at Strathclyde University (Milne, 1970e). Details of this enclosure are given in Chapter 8. After the rainbow trout fry had been hatched out in their hatchery in Stirling, Stirlingshire, they were moved to the sea site for growing to marketable size. Eleven tonnes of "marine" rainbow trout were harvested from Strom Loch in 1969, and this had risen to 16 tonnes by 1970.

The Highland Trout Company (Scotland) Ltd were the third concern to establish a "marine" rainbow trout farm, this time at Otter Ferry on the shores of Loch Fyne in Argyllshire. By damming the adjacent Largie burn a fresh water gravity supply was available, and pumps are only required for the sea water supply. The American technique of concrete raceway cultivation has been used at Otter Ferry. A bank of 20 of these raceways, 30 m long by 2·4 m wide and 1 m deep, have been built on a sloping hillside, as shown in Fig. 40, so that the water from raceways nos. 1 and 2 can be used by nos. 3 and 4. The floor is sloped down to the outlet with a fall of 7.5 cm in the 30 m length. The fresh water and sea water are mixed together in a common sump to give the required salinity for supplying to the raceways (Milne, 1970c). The young rainbow trout fry can therefore be transferred straight from the adjacent hatchery to the raceways in fresh water, and slowly acclimatised to sea water by gradually increasing the salinity. Work started on this farm in 1969, and in 1971 22 tonnes of rainbow trout were sold to the market. Plans for expansion are underway at present (December, 1971) for an annual production of 110 tonnes by 1972.

Rainbow Trout Culture in Australia

The cultivation of rainbow trout, *Salmo gairdnerii*, is still in its infancy in Australia due to two factors. The first is that until 1960 it was illegal to farm trout on a commercial basis, and second that the water flows and temperatures in Australia are generally unsuitable. However, Tasmania is an exception with more reasonable temperatures, and water flows suitable for trout farming. Legislation was passed in 1963 in Tasmania to permit farming there and the first licence issued to Sevrup Fisheries Pty in 1964 for commercial rainbow trout farming (Purves, 1968).

The area selected was Bridport on Tasmania's north-east coast. Although the Brid River is not large it provides an assured flow. The farm is built on a reclaimed tidal mudflat in the sea. Before entering the sea, the Brid River enters what was once the

Fig. 40 Bank of raceways, each 30 × 2·4 × 1 m (for rainbow trout and salmon farming) built on a sloping hillside at Otter Ferry, Loch Fyne in Scotland. (*Credit—P. H. Milne*)

estuary of a much larger river, but this was diverted during the first world war, and the estuary is now a tidal lagoon with a good supply of sea water.

The supply of rainbow trout eggs used at the farm came to Australia via New Zealand from California last century and with no imports since then, the stock supply is good. After hatching, the fry are kept in asbestos cement troughs until they are 40 mm long. Once they reach this length the fry are transferred to a concentric Foster-Lucas pond, for growing up to 15 cm. They are then ready for transfer to one of the nineteen growing ponds on the reclaimed foreshore. These are earth ponds built on the Danish principle (Sedgwick, 1966), and the sea water supply is pumped from the estuary. The fresh and salt water supplies can be regulated to provide the required salinity for the hatchery, rearing unit or ponds.

After successfully establishing the fresh water side of rainbow trout farming, Sevrup Fisheries are progressing into the development of sea water

rearing, and have recently purchased 20 hectares of land on the opposite side of the estuary for sea water expansion, using concrete raceways.

As well as the farming of table size fish, Sevrup Fisheries also propose to grow the rainbow trout to 2½ kg in sea water to compete with the massive imports of salmon, since there are no natural salmon in Australian waters.

Salmon Culture in Norway

Early research into the possibility of cultivating the Atlantic salmon, *Salmo salar*, was started by the Vik brothers at Vike-oyra in the summer of 1959 (Vik, 1963). By 1963 it had been shown that salmon could not only be kept in captivity, but that they could be made to spawn in successive years, provided they were changed from sea water to fresh water at the right time each year. From the start the hatchery techniques were studied and eggs, from the initial salmon kept in captivity, were hatched and reared up to the smolt stage. By 1962 some 40 of these salmon had reached maturity and produced 50,000 eggs, after gradually being accustomed to sea water and transferred back to fresh water again before spawning.

Large-scale commercial production of salmon has, however, been slow to develop due to the problems of the transition stages from salt to fresh water. One of the first firms to develop plants purely for salmon production was A/S Mowi, with four sites in the vicinity of Bergen. Two of these are for the breeding of smolts, dealt with in this section, and two sea water sites for the breeding of the salmon up to marketing size, as dealt with later in Chapter 8.

The brood stock salmon at A/S Mowi's two hatcheries at Oyerhamn and Tveitevågan, are stripped of their roe and milt in November and December, and blended together to produce fertilisation. After placing in special hatching troughs of running water, it takes till the spring for the salmon eggs to hatch and become yolk-sac fry. Just before the food supply in the sac is exhausted the fry are transferred to breeding tanks which are fed by automatic feeders. These breeding tanks are 3 m in diameter and 1·0 m deep, constructed from glass fibre reinforced polyester, Fig. 41. Both fresh water and salt water can be supplied to each tank, the former from a nearby lake, the latter pumped from the sea. The proportion of fresh/salt water can then be adjusted from fresh to salt depending on the growth of the fry. At the optimum production A/S Mowi reckon to raise 80% of the fry from parr to smolt in one year. In nature this process takes two to four years.

A/S Mowi's first hatchery is sited at Oyerhamn, on the island of Varaldsøy in the Hardangerfiord. This is the main production plant with 60 circular tanks for smolt, with a hatchery capacity of one million fry and an annual production of 200,000 smolts. To accelerate the process of hatching, the plant is equipped with facilities for heating the water to speed up production. A/S Mowi's second hatchery is at Tveitevågan on the island of Askøy near Bergen, and is responsible for both smolt production and brood-stock with a hatchery capacity of one million fry and an annual supply of 100,000 smolts. This plant has 10 circular tanks, Fig. 41, like Oyerhamn and in addition several raceways and large concrete tanks, Fig. 42. Fresh water is supplied by a nearby lake and salt water is pumped ashore. Research is also being carried out here into the controlled effect of heating the hatchery water to increase production. As well as the large number of brood-stock salmon kept here, several "wild" brood-stock are also caught to augment the egg supply. Once the smolts have been acclimatised to full sea water they are transferred

Fig. 41   3 m dia. smolt tanks at the Tveitevågan salmon hatchery in Norway.   (*Credit—by courtesy of A/S Mowi*)

Fig. 42 Smolt production and brood stock tanks for salmon at Tveitevågan in Norway. (*Credit—by courtesy of A/S Mowi*)

in special sea water containers to the two sea water enclosures, discussed in Chapter 8.

Salmon Culture in Scotland

The establishment of the commercial cultivation of the Atlantic salmon, *Salmo salar*, in Scotland, like Norway, has followed the successful research by the Vik brothers (Vik, 1963).

Salmon culture in Scotland has been developed in parallel with rainbow trout discussed earlier in this chapter, at two farms. The first, constructed by Marine Harvest Limited at Lochailort and the second by the Highland Trout Company (Scotland) Ltd at Otter Ferry, Loch Fyne. The onshore tanks, floating cages and general techniques of these two concerns are dealt with in the description of rainbow trout culture in Scotland, so there is no further need to describe the facilities since they are common to both species; the main requirement being tanks which can have the salinity varied from fresh to full sea water. However, some extra salmon tanks have been constructed at Otter Ferry, as shown in Fig. 25. Since these developments are in their infancy, production figures are as yet not meaningful.

Salmon Culture in United States of America

With the recent establishment of a Sea Grant Programme in the United States for the cultivation of marine finfish and shellfish, considerable advances have been made in hatchery techniques for Pacific

Fig. 43 Aerial view of fish spawning channel and fry rearing pond (marked with cross) at Wells Hydro-electric Project on the Columbia River. Both channel and pond are PVC-lined. (*Credit—by courtesy of Staff Industries*)

Fig. 44 Joining 20 m wide sections of PVC liner for the 3·6 ha fry rearing pond at Wells Hydro-electric Project. (*Credit—by courtesy of Staff Industries*)

Fig. 45 After laying the PVC liner at Wells Hydro-electric Project it was covered by 20 cm of gravel to simulate river conditions. (*Credit—by courtesy of Staff Industries*)

Fig. 46 Installation of vinyl-PVC liner in salmon rearing pond in Washington. Upper edge of liner anchored in soil as shown in Fig. 33. (*Credit—by courtesy of Staff Industries*)

salmon, *Oncorhynchus* spp., especially as a result of the construction of hatcheries to replace the lost natural conditions of salmon rivers that have been dammed for hydro-electric development (Edson *et al.*, 1971; Loder *et al.*, 1971), Figs. 43–46.

Initial studies by Oregon State University, under the Sea Grant Programme were concentrated on the chum salmon, *O. keta*, with the setting up of a hatchery to rear one million fry at Netarts Bay in the spring of 1970. In the first year 700,000 chum salmon fry were successfully reared using gravel incubation tanks (McNeil, 1971b). Other species to receive attention are the pink and Chinook salmon, *O. gorbuscha* and *O. tshawytscha*.

The most difficult task in salmon cultivation is the acclimatisation of the fry from fresh water to the smolt stage for transfer to salt water. This topic has been investigated at the Oregon State University Marine Research Laboratory at Port Orford, and Chinook salmon were successfully acclimatised to salt water in 30 days after yolk absorption with only a 4% mortality. Before transfer to salt water, salinity 33‰, the fry were kept for 20 days at 17–18‰ at a temperature of 11–12°C and fed heavily, in order to achieve initial rapid growth in water only slightly hypertonic to blood and tissue salts (McNeil, 1971a). However, it was discovered that the fish transferred to 33‰ exhibited slower growth than those retained at 18‰, with a difference between 2% increment in body weight per day compared to 3% increment in body weight per day. Additional research is now planned to determine at what age and size it is best to acclimatise juvenile Chinook salmon to sea water without significant impairment to growth.

The trend towards the use of ponds, bays and impoundments for the rearing of salmon is discussed in Chapters 7 and 10.

Pompano Culture in United States of America

Considerable interest has been generated in the cultivation of the pompano, *Trachinotus carolinus*, in Florida and other areas along the Gulf coast. Although several attempts have been made since 1957 in the commercial raising of pompano, all have so far foundered due to insufficient knowledge on the biology and ecology of the pompano (Berry and Iversen, 1967). Despite this lack of commercial production at present, since pompano are fast growing fish and command a high price, numerous governmental, academic and commercial organisations are engaged in pompano mariculture studies (Finucane, 1971b).

One of the major existing problems for the commercial raising of pompano is that no hatchery techniques are available for the production of pompano fry, and thus the work is dependent on wild stocks for the supply of fry and adults. Another problem is the availability and type of food required for feeding the pompano, both in the hatchery and in enclosures, up to marketable size.

As a result of the early unsuccessful attempts in raising pompano in ponds in Florida, as detailed in Chapter 7, considerable research has now been carried out, not only into the above problems, but also into the other related problems of environmental control. This recent work has been carried out mainly by two concerns, Tropical Atlantic Biological Laboratory of the Bureau of Commercial Fisheries and the Institute of Marine Science at the University of Miami, both in Miami, Florida.

Early research carried out for the Minorcan Seafood Company at Marineland near St Augustine on the experimental raising of pompano has been reported by Fielding (1966) and Moe *et al.* (1968), describing some laboratory tests to determine the salinity, temperature, dissolved oxygen and pH tolerances of pompano. The artificial spawning and embryology of the pompano has been studied by Finucane (1969, 1971a) in order to devise methods for the hatchery production of pompano fry from the egg to ensure an adequate egg supply for commercial concerns. Although the recent (1971) experiments were unsuccessful in producing fry from the eggs, considerable experience was gained in the development of techniques, and these appear to be more promising for the culture of larvae in the future.

Further research into the environmental tolerances on the growth of pompano has been carried out at Miami Seaquarium (Iversen and Berry, 1969), with six samples of pompano fry caught by beach seining in the surf zone. The average weight of these pompano increased almost 200 g over a 4–5 month period during the summer months. A decrease in this growth rate would be expected at lower winter temperatures, the absolute minimum being 10°C without mass mortalities. The results of the Miami Seaquarium tests gave better growth rates than those quoted by Moe *et al.* (1968), no doubt due to more consistent feeding, better water conditions and a more suitable environment than that of Marineland.

Tests are also being carried out into the suitability of shore ponds or sublittoral enclosures for the farming of pompano; the latter facilities are described in Chapter 8.

As mentioned earlier the water temperature is critical for the cultivation of pompano with the minimum at 10°C. Studies of growth have indicated that the best growth is obtained with water temperatures of 25–27°C (Finucane, 1971b). To obtain such high temperatures the University of Miami in conjunction with the Florida Power and Light Company at Turkey Point have developed a system of ponds, under the Sea Grant Programme, adjacent to the Turkey Point power plant, as discussed earlier for shrimp culture. The site covers 2 hectares, and includes one 0·4 ha, two 0·2 ha and four 0·1 ha ponds (Tabb *et al.*, 1969), Figs. 32–34.

Plaice Culture in Britain

In the late nineteenth century, several marine hatcheries were built in Europe to augment the fish stock of inshore waters. These early hatcheries were set up to exploit the spawning of plaice, *Pleuronectes platessa*, and cod in marine ponds, and once the eggs had hatched, the yolked larvae were liberated in the sea. However, since the mortality rate was still high during the larval stage, there was little measurable effect on the subsequent yield of marketable fish, and consequently many of the hatcheries closed down.

At the turn of the century experiments were carried out on the transplantation of young plaice from inshore waters to the Dogger Bank (Garstang, 1905; Borley, 1912) to try and increase food production. It was found that the average growth in weight of the transplanted fish amounted to 382% of the original weight, against similar fish in inshore grounds which only increased in weight by 100%. But these transplanted fish still had to be hunted for harvesting, and an enclosure technique was required.

One of the first British attempts at studying the possibility of the cultivation of marine fish, plaice, was organised by Gross (1944) in Loch Sween, Argyll in Scotland during the second world war. Gross and his colleagues first carried out experiments in the isolated Loch Craiglin, off Loch Sween (Gross *et al.*, 1949), which was enriched by the addition of fertilisers, sodium nitrate and superphosphate. A striking increase in the phytoplankton and other vegetation followed. The plaice released in the loch grew twice to four times as fast as those in nearby untreated waters. A second experiment took place in Kyle Scotnish, another arm of Loch Sween, and supported the earlier findings that flatfish feeding in an enriched loch grew very much faster than those in untreated adjacent areas (Gross *et al.*, 1952).

The main problem for the development of fish

farming was still the production of a satisfactory supply of hatchery stock beyond the larval stage. The prospects for the mass culture of marine fish improved substantially when Rollefsen (1939, 1940) in Norway found that the nauplius of the brine shrimp, *Artemia salina*, was an easily cultured and acceptable food for larval plaice. This was the breakthrough necessary for rearing marine fish.

Attempts to rear plaice under laboratory conditions began in the early 1950's at the Lowestoft Laboratory of the Ministry of Agriculture, Fisheries and Food (Simpson, 1959a,b; Shelbourne, 1953). The early experiments were carried out in tanks to study the larval growth rates and natural mortality curves. This work subsequently led to basic morphological studies on plaice from tanks and from the sea (Shelbourne, 1955, 1956a,b, 1957).

These studies were, however, carried out in a closed circulation system at Lowestoft (Shelbourne *et al.*, 1963; Riley *et al.*, 1963) and by 1960 it was evident that an accessible supply of plaice eggs and good quality sea water was essential for further progress. The Ministry of Agriculture, Fisheries and Food therefore carried out experiments in 1960 at Port Erin on the Isle of Man where these requirements could be met (Shelbourne, 1963). Rapid progress was made at Port Erin with increased survivals in 1961. Followed in 1962 by studies on the effect of bacterial control during egg incubation, and further the effect of temperature control whilst using antibiotics. By 1963 an efficient hatchery technique had evolved for the mass culture of plaice larvae to metamorphosis, when the symmetrical larvae flatten and settle on the sea bed (Shelbourne, 1964, 1965).

These successful experiments resulted in a joint programme between the Ministry of Agriculture, Fisheries and Food and the White Fish Authority for a hatchery at Port Erin in 1963 with a target output of one million plaice. During the first season of full-scale operation (1964), 160,000 young plaice were produced and by 1965 this output had risen to 500,000 plaice.

By this time the White Fish Authority were studying the feasibility of the marine fish farming of plaice up to marketable size. Consequently a 2 hectare intertidal enclosure at Ardtoe, Argyll in Scotland was chosen in 1965 for high density cultivation, as discussed in Chapter 7.

The experiments with plaice and sole, *Solea solea*, at the Port Erin hatchery had shown that at water temperatures higher than those found in nature, there was an increase in the growth rate of the fish. With a ready supply of heated water from

coastal power stations at about 7°C above ambient it was thought that this water might be utilised for fish farming. However, it was not known whether the fish could withstand the chlorine and other chemicals added to the effluent water to prevent the settlement of mussels and other organisms within the cooling water system.

Two pilot plants were therefore built, the first at the Central Electricity Generating Board's power station at Carmarthen in Wales and the second at the South of Scotland Electricity Board's nuclear power station at Hunterston in Ayrshire. In 1966 it was demonstrated at both these power stations that plaice and sole could survive, and also grow faster than fish in unheated water (Nash, 1968).

The success of these trials encouraged the White Fish Authority to construct a test facility at Hunterston to determine whether it was possible to grow fish to marketable size in the heated water effluent, and also to develop a commercially attractive cultivation technique. Originally four large tanks were constructed, $7.5 \times 15 \times 1$ m, and the facility provided with a supply of both heated and ambient sea water to allow the actual temperature in the tanks to be controlled (Nash, 1970). These four tanks are shown lying between the office block and the hatchery in Fig. 47. This aerial photograph taken in 1971 shows the growth and expansion of the unit since 1965.

The hatchery techniques which originated at

Fig. 47 Aerial view of the White Fish Authority's Fish Cultivation Unit at Hunterston Nuclear Power Station in Scotland. Bottom right is the office block, with indoor tanks in building on left. (*Credit—P. H. Milne*)

Fig. 48   Polythene tubs and drums used for hatchery tanks indoors at the Fish Cultivation Unit at Hunterston.
(*Credit—P. H. Milne*)

Fig. 49   Large indoor concrete and fibreglass sectional tanks for use in flatfish culture (plaice, sole and turbot), at Hunterston.   (*Credit—P. H. Milne*)

Port Erin have now been further developed at Hunterston using small polythene containers, Fig. 48. Indoor concrete ponds and fibreglass sectional tanks are also accommodated in the hatchery to provide controlled environmental conditions for the rearing of the fish, Fig. 49.

These experiments at Hunterston have shown that plaice and sole can be grown to marketable size from the egg in 18 months (Swift, 1968, 1969; Richardson, 1970, 1971).

Sole Culture in Britain

Investigations into the feasibility of rearing sole, *Solea solea*, in captivity have been studied simultaneously with that of plaice culture by the Ministry of Agriculture, Fisheries and Food and the White Fish Authority, as described above. After the initial experiments at the Port Erin hatchery, full-scale pilot trials have been carried out by the White Fish Authority at Hunterston power station in Scotland, Fig. 47, where sole have been reared from the egg to marketable size in 18 months (Richardson, 1971). The sole is in many ways a very suitable fish for sea farming since it is intrinsically valuable and the supply from the natural fishery seldom equals the demand (Cole, 1968).

The White Fish Authority experiments have all been conducted with an open circulation at Hunterston. However, sometimes the heated water supply from coastal power stations is of variable salinity, and carries silt in suspension.

To avoid these problems the Unilever Research Laboratory near Aberdeen have developed a re-circulation system with only 5% supplemental sea water. In the first year of operation, by developing self-cleaning tanks and an associated water treatment plant, Unilever were able to rear 40,000 Dover sole, *Solea* sp., beyond metamorphosis (Phillips, 1970).

The sole are hatched in conical incubators and the sole larvae left for two days to absorb part of their yolk sacs. They are then counted and stocked in 60 cm diameter cylindrical tanks at about 1 per ml and fed on brine shrimps, *Artemia salina*. The water depth in these tanks is varied as required and the downward passage of the water removes waste material, the outlet being screened by a nylon mesh floor.

The waste products, food, faeces and dead *Artemia*, are filtered from the outlet by a 30 cm deep bed of mussel shells overlying 120 cm of crumbled granite. Heating, ultraviolet sterilisation and aeration complete the water treatment process, which provides clear oxygen-saturated near-sterile water for reuse at 15°C.

Chapter 7
Intertidal Farming Techniques

Sea farming in the intertidal zone, between high and low water, takes advantage of the tidal range for water movement and circulation. To hold fish and crustaceans in this region impermeable barriers with sluice gate controls are required. Sessile organisms such as molluscs, however, do not necessarily need such costly constructions and can be attached to racks and frames in tidal seaways. The design and construction of intertidal farming techniques for sea farming are discussed with examples from many parts of the world.

The creation of a pond enclosure in the intertidal region entails the impoundment of sea water by a seawall, bund or barrage and enables the environment within that pond to be fully controlled. To maintain a marine environment under such conditions may require sea water circulation during each tidal cycle. If the pond has been properly designed from both the biological and engineering aspects it should be possible to utilise the daily variations in tidal range to effect the necessary sea water replenishment, without recourse to pumping. It is thus essential to ensure that biological requirements are met by the intertidal site, and this entails careful site selection as detailed in Chapter 4. If the pond is designed with the water level below the high water neap tide mark, and the bottom of the pond above the low water neap tide mark, as shown in Fig. 13, full control is possible. The pond can therefore be filled or emptied at any time during the tidal cycle to permit either circulation or harvesting. These requirements must therefore be fully considered in the initial design of the sluices to ensure satisfactory management during operation.

A reduction in salinity of the water in the pond due to rainfall must also be considered between each replenishment period. As mentioned in Chapter 4 a hydrological survey is essential during the design stage to ascertain rainfall expectations. The inflow of this fresh water from the pond's surface area and the surrounding catchment area must therefore be assessed to see if it would significantly alter the pond's salinity. If there are several streams from the hinterland flowing into the pond it might be necessary to divert these streams. One solution could be to provide a fresh water channel round the perimeter of the pond, draining into the sea. Whatever method is adopted, surface sluices or spillways for skimming off the surface fresh water due to rainfall are required in the design of the barriers.

Many fish farms have been constructed in the intertidal zone, especially in the Far East, for the pond culture of brackish water fish. The majority of these ponds are constructed from mud, and sometimes clay, on the intertidal mudflats near river mouths, and their design and construction are considered later, as the techniques used are very applicable to mariculture. Various other materials such as sand, stone, cement and man-made components are discussed for pond construction with reference to the various engineering techniques available.

The selection of materials for mollusc culture, and the different methods used in the intertidal zone throughout the world are described.

(I) MOLLUSCS

Oyster Culture in Japan

Japan is the largest oyster producing country in the world. Although seven species of edible oysters occur naturally in Japanese waters, only one is cultivated, the so-called Japanese oyster, *Crassostrea gigas*. This cultivation of oysters has been carried out in Japan since at least the eighteenth century. The two major areas for modern cultivation are firstly in Matsushima Bay, near Sendai on the north-east of the main island, and secondly at Hiroshima on the shores of the Inland Sea (Yonge, 1970). Only the first is discussed in this intertidal section, since the second uses floating raft techniques discussed in Chapter 10.

Suitable sites for the collection of oyster spat require to meet several conditions (Fujiya, 1970):

(i) The farming area must be protected by nature or by simple breakwaters against wind and wave action during storms;

(ii) Adult oysters, which produce the spawn, must be indigenous to the area to ensure adequate rates of reproduction;

(iii) The tidal current flow must be sufficient to change the water of the area completely and frequently;

(iv) The oceanographic conditions of the area, especially the salinity and water temperature, must be suitable for oyster growth;

(v) The water must contain adequate nourishment for plankton production and should contain suitable food for the oysters and larvae, such as phytoplankton;

(vi) The area must be protected against industrial wastes and sewage since molluscs concentrate metallic elements in their tissues (Fujiya, 1965); and

(vii) The water of the area should be clean from a sanitary point of view.

Matsushima Bay, near Sendai, was originally the centre of the seed-oyster industry in Japan, but this has now shifted further up the coast to the less heavily populated Ojika Peninsula. Here in shallow water the seed oysters are collected on rens attached to stakes driven into the bottom. These collecting rens consist of a 2 m length of No. 16 galvanised wire bearing either scallop or oyster shells. If the oyster seed is for local Japanese use, each ren has approximately 100 scallop shells as collectors, but if the seed is for export 70–80 oyster shells are used as the spat collectors.

These collecting rens are doubled over and suspended from horizontal bamboo poles set just below the low water spring tide level, so that they are not exposed to the air at any time, Fig. 50. These horizontal poles must therefore be at least 1 m off the bottom, and preferably placed at the edges of tidal streams. Two types of frames are used to support the horizontal poles. The first and more simple frame, consists of two crossed bamboo poles, like an ×, driven into the seabed at 1½–3 m intervals with the horizontal pole tied into the crotch. A second more elaborate framework consists of a rectangular bamboo platform 1 m wide by 3–6 m long which is supported on uprights driven into the seabed. The 2 m long collecting rens are then draped closely packed together over the horizontal bamboo poles (Ryther and Bardach, 1968).

In the Sendai region of northern Honshu, the peak of spawning is late August till September. During this spawning period, biologists at the various prefectural laboratories carefully monitor

Fig. 50 Bamboo frames for oyster culture in Japan. The 2 m long wire rens are doubled and hung over the horizontal bamboo poles. (*Credit—by courtesy of the White Fish Authority*)

the plankton and put out test collectors to forecast the settling peak for the oyster spat. The biologists also advise the oyster growers when to avoid heavy fouling from barnacles and other organisms. This co-operation between the laboratories and the growers ensures the best settlement of spat fall obtainable, since they are advised of the best time to put their spat collectors in the water.

After one month, when the seed oysters are 5–10 mm in diameter they are moved to hardening racks. This hardening process is necessary for the export of seed oysters, otherwise they would not stand being transported to the United States of America, their main importer. Initially at 5 mm the rens on the platform frames are laid horizontally along the top, so that they are just exposed at low water. Later at 10 mm they are raised higher still into the intertidal zone, until they are exposed for 4–5 hours during each tidal cycle. Sometimes the hardening is carried out in one stage.

The seed oyster industry in Japan requires some two million ren to produce on average, an annual production of 2,000 million seeds. About half of this production total is exported to the United States for cultivation; hence the care and attention given to the hardening process for export.

Oyster Culture in Brittany, France

The cultivation of the European or "flat" oyster, *Ostrea edulis*, is centred in the Brittany region of France. This is one of the oldest examples of mariculture, having been practised in many parts of Europe since Roman times. The oyster spat are mainly collected in the Gulf of Morbihan and then planted in specially prepared parcs in harbours and estuaries where they are exposed only during low water spring tides.

The time to place the spat collectors in the estuary is most important to the growers, firstly to obtain a good spat fall and secondly, to avoid fouling with barnacles and other organisms. The growers in this area are provided with a forecast of the peak times for spat setting by biologists employed by the Field Station of the Institut Scientifique et Technique des Pêches Maritimes at Carnac (on the Gulf of Morbihan) who carefully monitor the plankton during the spring and summer for the appearance of the mature oyster larvae (Ryther and Bardach, 1968).

The traditional spat collectors consist of semicylindrical ceramic tiles about 0·3 m long and 125 mm in diameter. The tiles are arranged in pairs, alternate layers being stacked at right angles to one another, with the concave surface downward. A stack consists of 5–6 pairs of tiles measuring 1 m high, wired together for ease of handling, Fig. 51. Before stacking, the tiles are covered in lime so that the young oysters can easily be flaked off in the spring and the tiles re-limed for the coming season (Yonge, 1970). These stacks of tiles are then placed on wooden platforms in the estuary raising them 15–25 cm off the bottom. After the spat fall, the collectors are left in position till the winter, when the seed oysters (approximately 50 per tile) are stripped off for transplanting to parcs in Brittany. The parcs or oyster beds are demarcated by palisades of stakes to discourage the entrance of fish which would consume the young oysters. These parc sites are chosen in areas where there is a rich growth of plankton usually in shallow areas

Fig. 51 Stack of semi-cylindrical ceramic tiles used as oyster spat collectors in France. (*Credit—by courtesy of World Fishing*)

with an annual temperature range of 5–20°C, salinity range 31–33‰ and a tidal range of 7–9 metres. The bays and estuaries in Brittany are chosen by the oyster industry for their protection from storms. These parcs are not necessarily the areas where the oyster larvae would settle as spat on the bottom, since in such shallow estuarine conditions the sites would soon become silted up without attention, so the French system is entirely artificial (Yonge, 1970). Nevertheless, this system has been highly successful with an annual production of 800 million oysters.

In recent years a new lightweight spat collector, consisting of a plastic mesh material (Netlon) made the same size and shape as the ceramic tile, has been used. These plastic spat collectors are stacked in a similar manner to the ceramic tiles but due to their light weight they can easily be swept away by tidal currents so it is recommended that two ceramic tiles are used at the bottom of the stack to act as ballast. The tiles before immersion are coated with a blended mixture of six parts of lime to one of cement to improve the adhesion of the oyster larvae. The removal of the oyster spat is more efficient with the plastic mesh since they are easier to detach by simply bending the tile whereas with the inflexible ceramic tile, 10% of the oyster spat is damaged on removal. Up until 1969 only 200,000 of the Netlon tiles had been sold, but as more experience is gained with their use this number is expected to increase (Nortene, 1969).

The extent of the industry can be gauged from the area of oyster beds in Brittany, which cover 5,200 ha. On the south coast there are about 500 individual parcs of 2 to 4 ha in size, whilst on the north coast there are approximately 200 parcs of about 20 ha each. The annual production of these oyster beds is approximately 17,000 tonnes.

Oyster Culture in Arcachon, France

The cultivation of the Portuguese oyster, *Crassostrea angulata* is centred in the Arcachon/Gironde district on the west coast of France (Labrid, 1969). Its introduction to the area was the result of a propitious accident in 1868 when a ship carrying seed oysters had to dump its cargo in the Gironde. These seeds established a natural bank of Portuguese oysters, which developed more rapidly than the flat oysters in the lower salinity and higher temperatures of the estuaries south of Brittany.

The culture of the Portuguese oyster in France is very similar to the European oyster already discussed. However, the Portuguese oyster is hardier and can withstand longer exposure to the sun and temperature extremes and thus may be placed higher up the beach. The traditional method of spat collection using ceramic tiles is now being replaced by wire mesh bags of clean oyster shells which are less expensive. Netlon, the plastic mesh described for use with the flat oyster tiles, is also readily applicable for forming mesh bags. These bags, approximately 50 cm × 100 cm, are then laid horizontally on wooden racks 15–25 cm off the bottom for the collection of spat.

Sometimes the oyster spat is left in the bags for two years until harvest time, but they grow faster if thinned out and placed on wooden racks or on the bottom. Although the Portuguese oysters do not require the hard, sandy bottom of the flat oyster, if laid on soft sediment, they require to be periodically raked to the surface to prevent suffocation.

Over 8,000 ha of intertidal beach is devoted to the cultivation of the Portuguese oyster in France with an annual production of 60,000 tonnes, giving a much better turnover than the flat oyster. Many hectares are also devoted to the "claire" culture of oysters in intertidal ponds.

In general the culture of oysters in France is carried out in the intertidal zone of the open sea with a natural food supply, discussed above. However, previous research into the use of agricultural fertilisers (Gross, 1952) had shown that the natural food supply could be increased by the addition of fertilisers but only in semi-enclosed locations, such as a pond, called claire culture, often using old salt ponds. Some French oyster ponds

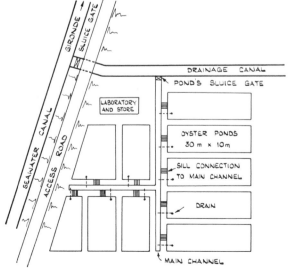

Fig. 52 Small French oyster ponds used for claire culture at Medoc.

Fig. 53   Medoc oyster ponds showing arterial channel and ponds dug out of brown clay.   (*Credit—P. H. Milne*)

constructed on the claire system were built at Medoc on the left bank of the Gironde in May 1964. The ponds lie 500 metres inland of a sea embankment, used to reclaim the land from the sea in the eighteenth century. Sea water supply comes from the drainage canals which are flooded at high water spring tides. However, like Audenge, the sea water can only be replenished six days per fortnight due to their operational level lying above the high water neap tide level. This small site of 10 ponds covering 0·4 ha is shown in Fig. 52 where the ponds are fed by an arterial channel. Individual ponds are 30 m × 10 m and vary in depth from ½ to 1 m due to differences in excavated pond bottom levels. The land on which the ponds are situated consists of a brown clay, Fig. 53, and part of the excavated material from the ponds was used to form the internal embankments; all the work being carried out by a small mechanical digger.

The preparation of the pond beds for oyster cultivation is a highly specialised job which requires to be carried out each year. The ponds are drained and allowed to bake and crack in the sun for about six weeks preparatory to receiving treatment. In order not to destroy the consistency of the clay wooden implements (rakes, spades and shovels)

are used in preference to steel bladed tools for levelling the pond. After preparation some of the ponds had blue superphosphate worked into the soil, and the results showed a doubling of the rate of growth of the oyster, not to mention an improvement in flavour as well. Various techniques are used for growing the oysters: some are spread over the bottom naturally, some retained in wire and plastic mesh boxes, 0·15 m off the bottom and some on 2 m long 1·5 mm diameter iron bars suspended 0·15 m above the pond bottom, Fig. 3.

The sluice gate for controlling the water supply was very simple, consisting of a reinforced cast iron plate and operated by a hand crank and rising spindle. Although the ponds are at different depths the sills for each of them are at the same levels and the clay sills are protected on the upstream side by 5 cm diameter wooden stakes driven vertically into the clay. The sluice gate scour channel was also treated in a similar fashion, Fig. 54. One problem of operation at Medoc is that all ponds are flooded to the same water level at the same time so individual pond control is impossible. The arterial supply channel is also used to drain away the less saline surface water on the following ebb tide when the sea water canal is empty. Any subsequent

Fig. 54 Sluice gate channel at Medoc oyster ponds showing scour protection using 5 cm diameter wooden stakes.

(Credit—P. H. Milne)

rainfall runs off into the arterial channel, and that can unfortunately reduce the incoming flood tide salinity. Although these ponds are of simple design at minimum cost, for pond management and salinity control it would have been better to have separate inlet and outlet channels, and individual pond sluices.

Although the maintenance of these claire ponds for fattening and "greening" the Portuguese oyster, *Crassostrea angulata*, is more expensive than open water culture, the oysters grow very much more quickly in the ponds, doubling their weight in a period of 4–6 months. Whereas the European oyster, *Ostrea edulis*, can only be stocked at a density of 4 per m², the same results can be obtained with Portuguese oysters stocked at 12 per m² (Ryther and Bardach, 1968).

Oyster Culture in the Philippines

In the Philippines oysters have been gathered from natural beds for many years, but since the second world war the natural oyster beds have dwindled.

In the Manila Bay area cultivation methods have in the past 30 years depended on stick and tray culture techniques due to the soft nature of the bottom sediment. The most common method is to drive bamboo poles into the seabed and Iversen (1968) reports that so many farms of this type are now established on the south-east shore of Manila Bay that it has become a forest of bamboo poles.

Several species of oyster occur in the Philippines, but the most important one for cultivation is the slipper oyster, *Crassostrea eradelie*. The simplest and commonest form of oyster culture consists of driving dried bamboo poles, 5–10 cm in diameter into the bottom at 0·3–0·6 m intervals. Originally oyster shells were wired to the poles as spat collectors, but just as many spat settled on the bamboo poles, so the bamboo poles are now used as the spat collectors. The poles are generally arranged in lines close together with just enough room to permit access by small boat.

In recent years biologists of the Philippines Fishery Commission have been trying to encourage more intensive culture techniques. One of these consists of a bamboo framework driven into the bottom. Each of the "plots" measures 1 × 10 m and is formed by three large bamboo horizontal members with 20 small diameter bamboo cross-pieces wired on at 30–60 cm intervals. This framework is held in position by a series of bamboo uprights driven 1 m into the bottom; the level of the framework being adjusted so that it is covered at high tide. From these bamboo cross-pieces, collector strings consisting of four or five oyster shells strung 15 cm apart, using No. 12 or No. 14 galvanised wire, are suspended into the water. Each "plot" is therefore capable of holding 140 collecting wires, each containing four or five shells. Only if the spat fall is excessively heavy is thinning necessary since the oysters grow very rapidly in both pole and plot culture and reach a marketable size of about 7·5 cm in 6–9 months (Ryther and Bardach, 1968).

The majority of the oysters in Manila Bay are cultivated by the pole technique, but the new hanging-culture platforms built under the direction of the Philippines Fishery Commission are being increasingly used, since by this method the yields are sufficiently greater to justify the construction of the framework. The density of these platforms is 300 per hectare, and each of these structures produces approximately 280 kg per year, representing a total yield of 82·5 tonnes per hectare per year. Since this yield is well in excess of the average yield for the pole culture areas in the Philippines, it does indicate the potential of intensive oyster culture in the region.

In addition to the cultivation of *C. eradelie*, recent tests have been carried out on the possible culture of the Japanese oyster, *C. gigas*. Seed oysters were imported from Hiroshima in Japan, and transplanted to the Naawan estuary in Mindanao. Over 67% of the oysters survived this transfer, and the rate of growth of the oyster is reported to be faster than in Japan, thus allowing the Philippine farmers a choice of species (Anon., 1968).

Oyster Culture in Australia

Oyster farming is Australia's oldest form of mariculture, and has been practised in New South Wales since the end of the last century. Commercial oyster production is limited to one species, the Sydney rock oyster *Crassostrea commercialis*. Initially cultivation techniques used stone or artificial shell beds, but the modern techniques developed since the 1930's employ tray and stick cultivation (Kestevan, 1941).

The Australian oyster farmer depends on natural spat fall for his stock. To collect the spat, sticks 2·5 cm square and 1·8 m long are nailed to frames 1 m wide, and secured to racks in areas known to be good catching areas for spat. The use of tarred hardwood sticks is considered to be the most suitable method of stick cultivation, mangrove sticks having been used originally. The 2·5 cm square size of stick is critical since thinner sticks are too flexible during wave action, and thicker sticks are too heavy for handling. These sticks are nailed into the 1 m wide frames at approximately 20 cm intervals ready for tarring. The sticks are then dipped in the tar for a few seconds, stacked to drain, and then allowed to dry for a few weeks before being taken to the catching area. Tarring sticks not only helps to protect the sticks from the ravages of borers and *Teredo* or shipworm, but also provides a clean and reasonably smooth surface to receive the spat fall (Ryther and Bardach, 1968).

These frames of sticks are then laid out in the catching area between the end of December and mid-January. Early placing of sticks is not recommended as they may become fouled with barnacles and other marine growths. The sticks are laid on racks consisting of two parallel, horizontal runners of 5 × 2·5 cm hardwood battens 1–1·2 m apart, held at the known catching level, about mid-tide, by posts driven into the bottom. These racks must be very firm and solid to withstand wave action and the mooring of small boats for maintenance. The sticks are then nailed or wired to the racks, and are

left there after spat fall until late winter or early spring.

Invariably the catching areas for spat fall are not ideal for maturing the oysters, and the transference of catching sticks, or "depoting", to maturing areas is very important in stick cultivation. There are several reasons for the depoting of newly caught oysters (Ryther and Bardach, 1968):

(i) it ensures that the sticks do not receive a second spat fall and thus all the oysters in one crop can be expected to mature at the same time;

(ii) the small oysters are protected from the attacks of predatory fish; and

(iii) the growth is forced and the young oyster spat develops quickly and becomes strongly enough shelled to withstand the attacks of fish when nailed out on growing racks without the necessity for the erection of wire netting fences against predatory fish.

The frames of sticks are generally left in the depoting areas until the following winter when the spat are approximately 15 months old. The nailing out on sticks in a maturing area generally commences in June or July and continues until late August. The racks to take these sticks in the maturing areas are designed exactly as for the catching area described earlier. These racks are constructed in parallel rows, which may be 5 m apart in good areas and 15 m apart in poorer growing areas. Each frame is nailed or wired to the rack, the latter method being more time consuming, but with less risk of dislodging any of the young oysters. These sticks are then left in position until the oysters reach marketable size in about three years from catching the spat fall.

It is often economic to transport the sticks long distances to the maturing areas if better growth rates can be achieved. One oyster farmer transfers his catching sticks from Port Stephens, 160 miles south to George's River in New South Wales to take advantage of more rapid growth to maturity (Iversen, 1968).

In addition to stick culture the use of trays has increased in recent years to produce a better product. After two years the oysters are transferred from the sticks to chicken wire trays and laid out on the racks as before for a further year before marketing (Iversen, 1968).

The annual product from the Australian oyster beds is 6,000 tonnes and the total area under cultivation is 6,600 ha. The average annual production is therefore 0·91 tonnes per hectare. However, the best areas under intensive cultivation produce 5·2 tonnes per hectare, using the tray method; and 2 tonnes per hectare, using the stick method. The economics of the extra labour in transferring the oysters to trays for the third year is thus apparent in the improved return.

Since the Australian oyster is grown in the intertidal region it is very susceptible to the extremes of winter frosts and summer heatwaves during low water. Particularly dangerous is the coincidence of extreme heat and low water during the day. To prevent oyster mortalities during these critical periods several of the oyster farmers have installed pumps on their leases to spray the exposed oysters with sea water on such occasions (Ryther and Bardach, 1968).

Oyster Culture in Canada

At the beginning of the twentieth century trial plantings of Pacific oyster seed, *Crassostrea gigas,* obtained from Japan were started in British Columbia (Quayle, 1969). A considerable oyster industry sprang up here as a result of the success of these trials, which was dependent on the regular importation of seed from Japan. Fortunately some of the plantings were allowed to mature and produce local spatfalls to ensure future breeding grounds.

The oyster spat is collected on old shells suspended from log rafts in the spawning areas. After settlement, the young oysters are removed from the shell and placed directly on the bottom in the intertidal zone. A farmer normally leases about 16 hectares, but much larger farms, over 100 hectares do occur. Once the oysters reach market size they are harvested by hand using rakes and forks.

It was not until the 1930's that oyster farming spread to eastern Canada. Conditions here are suitable for the American oyster, *Crassostrea virginica* (Galtsoff, 1964).

The growing season is, however, short, and it requires from five to seven years for the oyster spat to reach the minimum harvestable size. Originally the oyster spat was collected on cardboard collectors, but recently tests have been conducted on a number of commercial oyster spat collectors. One of them was made of polyvinyl chloride pipe frame with plastic mesh bottom and sides, suspended from long-lines or floats (Ruggles, 1969). An analysis of the spat settlement showed this method to compare very favourably with the standard trays, especially since the cost is approximately halved. Due to the long time from laying the oysters to harvest time, oyster farming in eastern Canada is usually a part-time occupation. Recent developments in the off-bottom culture of oysters in Nova Scotia appear more promising, as described in Chapter 10.

Oyster Culture in Cuba

Experiments started in Cuba in 1965 to investigate the cultivation of the mangrove oyster, *Crassostrea rhizophorae,* and these were carried out by the Centro de Investigaciones Pesqueras of the Instituto Nacional de la Pesca (Nikolic, 1970).

At the outset several different types of collectors for the oyster spat were tried ranging from branches, sticks and roots of different bushes, shell strings, shell in baskets or cages of galvanised or plastic wire, tiles of fibrocement and concrete plates. The best settlements of spat occurred on the terminal branches of red mangrove. These branches were hung from horizontal poles of palm and bamboo, supported by stakes driven into the seabed in the intertidal zone. The average annual yield per branch collector was 370 oysters, and the first harvest of commercial-sized oysters was possible after 4–5 months.

This work was supervised by the FAO, who are now training technicians for commercial scale operations.

Oyster Culture in New Zealand

Commercial oyster cultivation in New Zealand centres mainly on the rock oyster, *Crassostrea glomerata.* The same stick technique is used for the collection of spat as described earlier for Australia, but it is more common in New Zealand to transfer the oysters to the tray method of cultivation for growing up to market size since this improves the production yield. Oyster cultivation is still in its infancy in New Zealand with a production from cultivation of 230 tonnes in 1968 (Sorensen, 1970). With the development of a new Marine Farming Act in 1968 coastal aquaculture is developing at a fairly rapid rate, and a large number of new leases have recently been granted for rock oyster farming. These stem mainly from the success of the oyster farming techniques in the governmental demonstration farms (Anon., 1971a).

Mussel Culture in France

The cultivation of mussels, *Mytilus edulis,* has been carried on for many centuries in south-west France, but only in the last century in Brittany. The reason for this late development in Brittany is due to there being no natural mussel beds, which means growers have to import the seed from the south. The main technique used in France is the "bouchot" system, using stakes in the intertidal zone.

The selection of areas for mussel spat collection is most important since young mussels are remarkably sensitive to light and require a relatively high light intensity in the region where they settle (Ryther and Bardach, 1968). This explains their natural distribution in the intertidal zone. Mussels grow best if permanently immersed and hence the floating raft culture in Spain, discussed later. In France, however, the "bouchot" system is used where 15–20 cm oak stakes, approximately 3 m long, are driven 1 m into the seabed in the intertidal zone so that they are uncovered at low water spring tides. Lines of these stakes are placed at right angles to the coastline, 35 cm apart, with sufficient space between the rows for access by boat and this empirical method of spacing appears to give maximum growth. The bottom 25 cm of each of the stakes is covered with clear plastic sheeting to prevent crabs and other predators attacking the mussels.

The collection of mussel spat starts in May–July and sometimes continues through to August and September. The spat are collected on pine collector "bouchots", or by suspending loosely woven 125 mm diameter ropes, 3 m long, in the intertidal region near a natural mussel bed. The young mussels tend to attach themselves by their byssus thread to the strands of rope, since mussels prefer to attach to fibrous or thread-like material. After settlement the spat on the collector ropes or "bouchots" are taken to the growing area. Clumps of spat are hung in bags of fine mesh cotton netting which rots on the "bouchots", or the ropes are wrapped in a spiral round the oak stakes. It is most important in the transplanting of mussels that they are cultured under similar conditions to their settling region, especially with reference to light intensity, temperature and salinity, otherwise there is a danger of detachment and migration.

As growth proceeds the mussels cover the complete "bouchot" and start falling off due to their weight, unless the outer layer is broken off and placed in string bags for transferring to new "bouchots". To prevent loss of mussels due to detachment it is necessary to carry out this procedure two to three times a year, up until market time, approximately one year from spat collection when they are about 5 cm long. If larger specimens are required they are left till the autumn, but no longer. Much of the cultivation work is carried out on the mud flats during low water, using small flat-bottomed boats, or "acons", which can be rapidly propelled over the mud, by means of a foot over the side at low water, and paddled when afloat (Mason, 1971).

Some growers tie branches of chestnut trees round the stakes to prevent loss of mussels, and this is termed "catinage". A simpler method is now available using large mesh Netlon (Nortene, 1969). Special Netlon "catinage", sheath nets are now produced which will expand to 1·40 metres circumference, and it has been proved that using this technique the production of one stake can be increased from 10 to 25 kg. Bag cords of Netlon are also available for transfer of mussels to new stakes or frames and there is said to be a smaller loss of mussels due to detachment.

The total mussel production in France amounts to 30,000 tonnes per annum. No figures are available for the total area under cultivation, but Ryther and Bardach (1968) estimate production at 5 tonnes per hectare (i.e. approximately 6,000 ha).

Mussel Culture in Italy

The cultivation of mussels in Italy follows the traditional "bouchot" system described for France, except here the technique is called "pergolari". The mussel spat, *Mytilus galloprovincialis* (the Mediterranean mussel), is collected on ropes ("rope seed") from good settlement areas and transferred to growing regions. These growing structures consist of wooden posts driven into the seabed at the low water spring tide mark, and the ropes are then attached to the posts. When transferred in the spring the spat measure about 10 mm long, and by the following autumn, 18 months after settlement, have reached market size of 70 mm.

Very little information is available on the total areas under mussel culture in Italy but Favretto (1968) reckons that 58 hectares in the Gulf of Trieste are under cultivation producing up to 4 tonnes per hectare per annum. He considers these figures low for the traditional culture methods, and suggests more rational and efficient systems should be used to take optimum advantage of the commercial possibilities for culture.

(II) CRUSTACEANS

Shrimp Culture in Singapore

The pond culture of shrimps in Singapore has developed over the past 30 years and covers both the genus *Penaeus* and *Metapenaeus*. Since the ponds are not artificially stocked, the industry relies on the young post-larval shrimp being carried into the ponds on the flood tide. The main species harvested from these ponds are *P. indicus*, *P. merguiensis*, *P. monodon*, *P. semisulcatus*, *M.*

burkenroadi, M. ensis, M. mastersii, and *M. brevicornis*. The method of culture is essentially simple in that the ponds are constructed in the intertidal zone, with sluices opened daily to admit the flood tide twice a day. Excess water then runs off during the ebb tide and the ponds' water level regulated to give a minimum water depth of 0·6 m. Various aspects of this method of shrimp culture have been previously described by Burdon (1950), Tham (1955), Hall (1962) and Kow (1969).

The shrimp ponds in Singapore are usually constructed on the site of mangrove swamps adjacent to tidal creeks and estuaries, and are selected only after considering the location in relation to the following factors:

(i) The presence of commercially important species must be assessed in the economics of shrimp culture since the indigenous species *P. merguiensis* has the highest market value, followed by *P. indicus, P. monodon, P. semisulcatus* and *M. brevicornis*;

(ii) The tidal range also assumes great importance here since the pond operators would prefer to harvest the shrimp from the ponds daily. In Singapore the average spring tidal range is 3·35 m, so any swamp site with a level less than 1·2 m above L.W.M.O.S.T. could be considered suitable. Such a site would give a tidal head of 0·6 m to 0·9 m to generate daily interchange in the ponds. With this method of culture both the shrimps and their food are brought into the pond during the daily tidal flow. Sites with levels greater than 1·2 m above L.W.M.O.S.T. are not considered satisfactory due to lack of replenishment and food supply over the neap tide period;

(iii) The location of the site in relation to the coast is also most important since the nearer to the coast the greater number of post-larval shrimps will be obtained. However, on the other hand the site should be neither too close to the coast to suffer wave and storm damage, nor too far inland where there is any danger of flooding with rain-water during the monsoon;

(iv) The texture of the bottom material for the pond must also be considered for the habitat of the shrimp. Hard clay is preferred for the bottom (as well as construction), and there should not be more than 50 mm of silt above the clay, since deeper deposits of silt become anaerobic with a subsequent decline in the shrimp harvest;

(v) The size of the swamp capable of development must also be appraised. The smallest economical size of pond is about 12 ha since the income derived from it is sufficient to support two fishermen to open and shut the sluices, etc., and still leave a profit for the owner of the pond. Smaller ponds than this still require two workers for operation and are thus not economical;

(vi) Swamps which have large fresh water streams draining into them are also not favoured due to the inundation of fresh water lowering the salinity during the monsoon. If other conditions are favourable, steps might be taken to divert the streams provided it could be done easily and cheaply; and

(vii) The last important factor in site selection is the swamp salinity, which should be in the range 24 to 30‰. However, this may not be possible, and in general 18‰ is taken as the lower limit for site suitability, provided the other considerations are met.

After careful site selection, the construction of the pond has to be carefully planned to ensure successful management. As mentioned above the ponds are constructed as close to the coast as possible, but to minimise storm damage, a fringe of at least 15 m of mangrove is left undisturbed between the coast and the site. The first operation is to clear the desired area of mangroves and level the site. The surrounding embankments are then constructed by removing clay from the pond bottom in slabs approximately 30 cm long, 15 cm wide and 10 cm thick. These slabs are then laid side by side to form the embankments, and each layer is allowed to bake in the sun before placing the next layer. This technique reduces the settlement of the embankments which, even when well constructed, can settle as much as 30 cm in a year. Silty clay soils are not recommended since they may settle by as much as 60 to 90 cm a year with continual maintenance. These embankments should be built high enough to prevent overtopping and wide enough to allow access for maintenance. The design of the embankment should also allow for the differential water pressure on each tidal cycle.

To ensure ample replenishment there must also be a sufficient number of sluices to allow the flood tide to flow smoothly into the ponds without any turbulence since it is believed this may scare the young shrimps from entering the ponds. To simplify the operation of the sluices in Singapore, they are generally built in batteries wherever possible, having concrete foundations and sides, using wooden stoplogs. The sluices are of sufficient size to allow complete filling or draining on one tidal cycle. The embankments are protected by wooden boards at the water line in the vicinity of these sluices to prevent scour by the daily flow of water. From the sluice to the pond the channels are preferred to be straight and they must be kept clear of plants and obstructions. To assist in the dispersion of the inflowing water to all parts of the pond, channels are provided in the pond bottom radiating from the sluice gate to the farthest corners of the pond. When the pond is excavated the bottom is sloped down towards the sluice gate and this enables the fishermen to concentrate the shrimps for harvesting.

The technique of operating these sluices is to open them when the flood tide is just more than 0·6 m above the floor of the sluice and this ensures a head of water which prevents the young shrimps swimming out of the pond. Once high water is reached in the day time a 13 cm square mesh wire screen is placed across the sluice to prevent the escape of shrimps on the ebb tide. Since the shrimps are buried in the sediment during the day, no harvesting can be carried out. At night, after high water is reached, the wire screen is replaced by a 10 m long fish net with a mesh size varying from 21 mm at the mouth to 9·5 mm at the cod end. The mouth of this net is fitted to a wooden frame which is placed in the sluice just before the tide ebbs. At low water the net is removed and the catch retrieved from the cod end for harvesting. In this manner continuous harvesting can be carried out over 20 to 22 days per month. Due to the constant replenishment of the pond water twice a day there is no fertilisation in the ponds. However, to increase the rate of production of plankton and other micro-organisms it is customary to drain the pond completely once a year and allow the sun to bake the soil for three to four days before refilling. It is believed that this procedure increases the shrimp harvest.

Shrimp Culture in the Philippines

The cultivation of the shrimp *Penaeus monodon* has been carried out for many years in the Philippines, but only as a secondary product in association with *Chanos chanos* (milkfish). In the last ten years there has been a steady exploitation of the shrimp fishery, and as *P. monodon* has a higher market value than the milkfish, ponds are now being stocked exclusively with shrimps (Caces-Borja and Rasalan, 1969).

In previous decades the post-larval shrimp fry were carried into the ponds with the flood tide and as in Singapore burrowed in the bottom until harvest time (Villadolid and Villaluz, 1951). However, now that there is a regular supply of commercially available shrimp fry caught along the shores of Manila Bay the ponds are stocked direct.

The selection of a pond site in the Philippines is restricted due to the small spring tide range of 1·5 m and a neap tide range of 0·3 m. The three main factors to be considered in site surveys are:

(i) It is desirable to have available sources of fresh, brackish and saline water for *P. monodon* cultivation. The reason for this is that fresh water is more favourable for development of the shrimp in the early post-larval stages, whilst brackish and saline waters increase the growth in the adult stages;

(ii) A sandy clay to clay loam type of soil is preferable both for construction and a good growth of algal food. The sandy clay also suits the burrowing habit of the shrimps; and

(iii) The area selected for the pond should be capable of flooding to a minimum depth of 0·9 m during the spring tide period, because replenishment is not possible at neap tides.

Pond construction in the coastal marshes and swamps of the Philippines is similar to that used in Singapore, and the main area is enclosed by a clay embankment built from the excavated pond bottom. The shrimp ponds vary in size from a few to over a hundred hectares. The model layout of a 10 ha *Penaeus* pond, designed by Delmendo and Rabanal (1956), shown in Fig. 55, gives a good idea of how these ponds are designed and operated. The main water control sluice gate leads to a head pond at the deepest portion, and this head pond supplies water to the two nursery ponds, each 0·5 ha, and to the two main rearing ponds, each 4·5 ha. Radiating channels are dug in the pond bottom to disperse the inflowing water, and also assist in concentrating the shrimps at harvest time.

The nursery ponds are prepared for stocking in the dry season when they are completely drained, and the soil allowed to bake for two to three weeks. Commercial N:P:K fertiliser is then added at the rate of 10 gm per sq. m. This serves to absorb excess carbon dioxide in the water, as well as supply the necessary calcium needed by the shrimps, especially during their moulting periods. The pond is

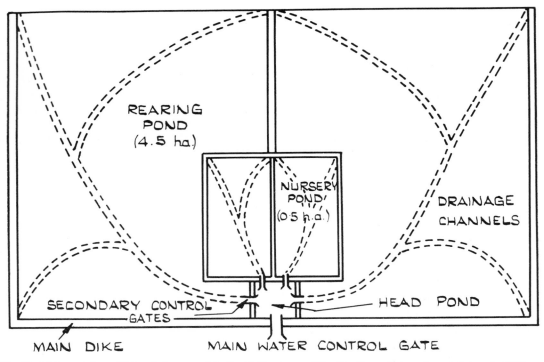

Fig. 55　Model layout for a 10 ha shrimp pond in the Philippines. (*Credit—From Delmendo and Rabanal, 1956*)

then flooded to a depth of 0·3 m and allowed to stand for another two to three weeks for the development of "lab-lab", a biological association of minute plants and animals growing on the mud floor of the ponds, which is the food of the shrimp. The pond water depth is then increased to 0·6 m and the ponds stocked at a density of 300,000 to 500,000 fry per hectare. The young fry are kept in the nursery ponds for four to six weeks and receive little attention apart from some occasional supplementary feeding with rice bran. Once they reach a length of 25 mm they are ready for transfer to the main rearing ponds. If the nursery ponds are adjacent to the main ponds as shown in Fig. 55, the fry can be driven directly through the sluices in the embankment into the main ponds. If the ponds are not adjacent the fry are caught in shrimp traps for transfer to the rearing ponds. The ordinary practice of the pond owners is to stock the rearing ponds at 1 per sq. m or 10,000 per ha, but if the pond has a good growth of "lab-lab" or supplementary feeding this may be raised to 2 per sq. m, or 20,000 per ha.

Due to the small tidal range in the Philippines, the water in the 0·6 m deep ponds is not changed regularly or routinely, but only as often as may be

necessary to maintain a temperature of approximately 25°C and a salinity of 20–25‰. The water may therefore have to be changed as often as once every two to three days in the rainy season, when the salinity may drop as low as 10‰, and in the hot, dry season when it may increase through evaporation to 35‰ with a temperature of over 30°C (Ryther and Bardach, 1968). With only one main sluice gate per pond only half the water is exchanged in the ponds at any one time. During replenishment screens of abaca cloth are placed in the sluices to both the nursery and rearing ponds to prevent the admission of predators and the loss of shrimps. After four to six months in the rearing ponds the shrimps have grown to a marketable size of 10–12 cm. To harvest the shrimps, the pond is generally slowly drained to concentrate the shrimps in the radiating channels down to the sluice gate. However, if only a few shrimps are required for market a variety of techniques can be used to trap or net the shrimps. One of the major problems in this type of shrimp culture is the difficulty of getting the shrimps out of their burrows when draining the pond, and if this is not carried out carefully, can result in the loss of those which remain buried in the mud.

Shrimp Culture in Japan

The development of hatchery techniques for the cultivation of the Kuruma shrimp *Penaeus japonicus* since 1933 have already been discussed in Chapter 6. However, the intertidal pond culture of these shrimps has existed since the eighteen nineties. This previous pond culture, using obsolete salt pans, was not so much production, but rather a holding method to take advantage of the seasonal price margin (Hudinaga and Miyamura, 1962). As the Kuruma shrimp has always been one of the most highly valued of Japan's great variety of seafood, with its demand exceeding the supply, there has been incentive into its cultivation.

When the Takamatsu Prawn Farm was established by Dr Fujinaga of the Kuruma Shrimp Farm Co. Ltd, a sea embankment was constructed in the intertidal zone. The sea water supply to the onshore tanks was originally impounded in a 1 m deep storage reservoir 4 ha in area, and the

Fig. 56 Pond aeration supplied by electrically driven paddle wheels in the main shrimp enclosure at Takamatsu.
(*Credit—J. D. M. Gordon*)

water supplied by gravity. To ensure adequate aeration four large electrically driven paddle wheels are used to churn up the surface and circulate the water, Fig. 56. As detailed in Chapter 6 this storage reservoir is now used as the main rearing facility for the shrimps once they reach 2 cm. The sea water supply is tidal and flows through two 0·6 m diameter pipes allowing filling and draining. Normally about 25% of the volume is exchanged on one tidal cycle, but the whole volume may be changed at spring tides if unfavourable anaerobic conditions occur. The water entering and leaving the enclosure flows through both large and small mesh screens to prevent the entry of predators and the escape of shrimps.

Dr Fujinaga's other prawn farm is at Aio on the island of Honshu. The site at Aio was formerly a salt-making project which was abandoned. Here two large ponds, each 4 ha, and several small ponds are open to the Inland Sea which pours through tide gates set in the seawalls. Adequate circulation is maintained by these tide gates without recourse to pumping which is always expensive. To harvest the shrimps from these ponds a special net is towed behind a motorboat. A jet of water, delivered under pressure from a pipe dragged over the bottom, dislodges the shrimp from the sand in front of the trawl net for harvesting (Idyll, 1965). The total annual production from the Japanese shrimp ponds amounts to 400 tonnes.

Shrimp Culture in the United States of America

At the same time as research in the U.S.A. into hatchery techniques for large scale cultivation of post larval shrimp were being developed (as detailed in Chapter 6), pond culture experiments were being undertaken to determine economic pond sizes and methods of construction (Webber, 1970). With an increase in the interest in the pond culture of shrimps, research ponds have been built in Louisiana (Broom, 1969), Texas (Wheeler, 1966) and Florida (Idyll *et al.*, 1969).

The most extensive research into pond design in the U.S.A. has been carried out in Louisiana (Louisiana Wild Life and Fisheries Commission, 1968; Broom, 1969) and much experience has been gained with the problems of maintenance and materials. Both the brown shrimp, *Penaeus aztecus* and the white shrimp, *Penaeus setiferus*, were used in the experiments. The first pond covering 0·1 ha was constructed in 1962 on Grand Terre Island, Louisiana. At the site, 1 m high levees were built from the silt and clay on the intertidal shore, but these could not be maintained because of wave action inside, and tidal action outside the pond. To rectify this, wooden bulkheads were constructed on both sides of the levee, but even this was unsatisfactory as the sand fill was washed through the joins of the bulkhead.

A second series of five 0·1 ha ponds, designated A-ponds, were built in 1964 at the same site, but this time using corrugated asbestos sheets as bulkheads, on either side of a 3 m wide levee. The asbestos sheets (3·65 m long by 1 m wide) were laid with the corrugations horizontal and were overlapped by 0·3 m. Each sheet was held in place by four 3 m long by 125 mm butt creosoted wooden posts, driven into the sand. These bulkheads were constructed 3 m apart and connected together with a 6 mm wire cable, and the space between them filled with sand from the pond bottom. A deep catch basin, as discussed for the Philippines, measuring 1 m wide by 2·5 m long by 150 mm deep was excavated at the deep end of each pond to assist in the harvesting of the shrimp, and the pond bottoms sloped in general from 0·8 m at the shallow end to 1 m at the deep end. The sea water intake and standpipe outfalls for these ponds were fashioned from 100 mm PVC pipe, and the bottom level of the catch basin was constructed at +0·3 m MLW to allow complete draining. These ponds were better than the original pond but problems still existed since sand often leaked out between the asbestos sheets, and the cables connecting the bulkheads kept rusting and required replacement.

These faults were rectified in the design of 16 0·1 ha ponds, designated D-ponds, completed in June 1968. These ponds showed several improvements over the A-ponds in that two layers of asbestos sheeting were used and were butted on end rather than overlapped. Square wooden posts were used in the D-ponds and placed at each point where the sheets butted, and in place of cables 188 mm, galvanised steel rods were used to hold the bulkheads together. As before, the ponds were designed to slope from 0·8 m at the shallow end to 1 m at the deep end, but the catch basin was enlarged to 2 m wide by 4 m long by 0·45 m deep and constructed in concrete with side slots for screens. To improve the drainage the bottom level of the pond was raised to +0·6 m MLW and the discharge standpipe increased to 150 mm.

During harvesting the improvements to the D-ponds were most significant since due to their higher elevation they could be drained completely and harvesting time averaged three man-hours. In contrast, the A-ponds required a small seine to harvest the ponds and also a pump to complete the draining with an average harvest time of 24 man-hours.

Supplemental feeding was necessary in the ponds for the shrimp to grow and the feeds used in the experiments were yellow corn meal, mullet and Purina Catfish Chow. Care must, however, be taken to ensure that the supplemental food is all utilised since overfeeding can result in anaerobic conditions with consequent loss of stock. These experiments showed that with care, densities of 300,000 shrimps per hectare could be managed without substantial mortalities.

Ponds at Louisiana stocked with unfed shrimps produced high mortalities, even at low density stocking, and the survivors consisted of small shrimps of poor quality. These conclusions were also arrived at by Wheeler (1966) who conducted research in Texas into the feasibility of cultivating shrimps commercially in ponds under semi-natural conditions. Two ponds were constructed, each 30 m by 15 m by 1½ m deep (i.e., 0·045 ha), and a different rearing technique was attempted in each. One pond (circulating-water pond) had a continuous exchange of sea water filtered through oyster shell, and the shrimps were fed daily with a prepared diet. In the other pond (static-water pond), water was only added to compensate for evaporation, and commercial fertiliser was added as in the Philippines to promote the growth of plankton, which is the natural food of the shrimp.

Both ponds were stocked in 1965 with 9,000 post larval brown shrimps, *P. aztecus* from the Galveston estuary. In the static-water pond a lush growth of plankton was obtained in the first three weeks after stocking. However, attempts to catch shrimp from the pond were unsuccessful, and on draining, the pond bottom was completely anaerobic with a complete shrimp mortality due to the low dissolved oxygen content of the pond. The circulating-water pond was, however, much more successful and the shrimps in it grew considerably during the experiment, emphasising the need for a continuous exchange of well aerated sea water to prevent de-oxygenation, as already commented on in the Japanese summary of shrimp culture.

Shrimp Culture in Indonesia

Recent successful experiments conducted firstly on the culture of the shrimp, *Penaeus monodon*, in brackish water ponds, and secondly on the artificial breeding of *P. merguiensis*, *Metapenaeus monoceros*, and *M. brevicornis*, have encouraged the Government of Indonesia to build a shrimp propogation centre in Makassar (Djajadiredja, 1970). The juveniles from this centre will then be distributed to farmers. An area of 10,000 ha of brackish water ponds in the province are expected to be used for shrimp culture.

The *P. monodon* used for the pond experiments were collected as post-larvae from the coastline and reared for 2–3 weeks in fine mesh net tanks, 1 m × 2 m × 0·6 m, stocked at 5,000 per tank. During this period the shrimps were fed finely ground fresh fish in the morning and benthic algae in the afternoon at the rate of 50% of body weight. The stocking ponds are fertilised to produce a bloom of benthic algae, as discussed earlier in shrimp culture in the Philippines, and stocked at a density of

30,000 per hectare. These shrimps reach marketable size in 6–7 months.

Shrimp Culture in Australia

1970 saw the construction of a large research station at Port Stephens north of Sydney, New South Wales, for studies on the feasibility of farming prawns in Australia (Anon., 1970e). The station has been built on a 12 ha reclaimed site in the heart of a swampy backwater. Water for the site is obtained from a short canal linked with Wallis Creek.

The main prawn which will be studied is the greasyback, *Metapenaeus bennettae*, which is an estuary dweller and therefore is most suited to this site. Research is also to be carried out on the king prawn, *Penaeus plebejus*, and the school prawn, *M. macleayi*, which are both of commercial potential, but little knowledge is as yet known about suitable densities for their culture.

In addition to the modern laboratory and aquarium facilities the site includes ten tanks and ponds (Anon., 1971j). Six of these are concrete block rearing tanks with a capacity of 110,000 litres, Fig. 57. The other four rearing ponds cover 0·1 ha and were excavated by grab line from the reclaimed swamp. The tidal range at the site is utilised to give a natural water supply by the construction of dams and sluices linked to the canal to Wallis Creek.

Turtle Culture in British West Indies

The culture of the green sea turtle, *Chalonia mydias*, was started in 1968 on Grand Cayman Island in the British West Indies. This development was established by Mariculture Ltd, using the techniques developed by Dr R. E. Schroeder, formerly a marine biologist with the University of Miami's Institute of Marine Sciences (Anon., 1971k).

In 1971 Mariculture were holding some 50,000 green sea turtles in their tanks and ponds. These turtles were grown from eggs collected from nearby sites where they had been laid by wild turtles. The eggs are then placed in an incubator for hatching, and afterwards are placed in 1·2 m × 0·9 m × 0·3 m wooden tanks and fed on a specially formulated high-protein pellet, which is similar to the diet of plankton which the baby turtles normally feed on for the first 6–8 months. They are later transferred to circular concrete tanks ranging from 6–15 m in diameter and 2·4 m deep, where they are fed a mixture of freshly harvested sea grass and pellets. The weights of the turtles average from 2 kg at six months up to 45 kg at three years. The optimum size for harvesting is 60–80 kg. Although each turtle only provides about 20% of its weight in turtle steaks the remainder is also marketable: shell for jewelry and ornamentation, bones and calipee for green turtle soup, skin for turtle leather goods and low-iodine, oxidisation-resistant oil for cosmetics (Anon., 1971k). The production target

Fig. 57 Port Stephens research station in Australia showing concrete block prawn rearing tanks. (*Credit—by courtesy of World Fishing*)

from the Grand Cayman turtle farm is 680 tonnes of turtle meat a year from a stock of approximately 60,000 turtles.

The present operation depends on available stocks of wild turtle eggs, but recently Mariculture have started to develop their own breeding stock of mature turtles, weighing up to 180 kg. These turtles are kept in a specially constructed "natural" pond, 100 m long by 50 m wide. The pond is 8 m deep at one side sloping up to a broad sandy beach to simulate their natural habitat for spawning. To provide sufficient water circulation a pump supplies 0·9 m³/s of sea water to the farm complex which is protected from the sea by a high seawall.

Turtle Culture in Australia

Following the successful establishment of a turtle farm in the British West Indies, Australian government scientists are carrying out research into the possible setting up of turtle farms in Torres Strait (Anon., 1971b). The object is to protect the newly hatched turtles during their first year, when, out of 600–800 eggs laid by a turtle each season, only two or three survive maturity. Dr Bustard in charge of the project hopes to increase this survival rate 50–100 times.

Three out of six pilot farms will be set up on Darnley Island and will be run by Torres Strait islanders and local Aborigines. The demand for

turtle products, oil for cosmetics, leather as a substitute for crocodile leather, the calipee cartilage for turtle soup, in addition to turtle meat, is encouraging investors, since turtles weigh up to 80 kg at harvest time.

(III) FISH

Pond Culture in Far East

Mud and clay have been the traditional materials throughout the centuries for the construction of fish ponds in the Far East (Schuster, 1949), for the cultivation of *Chanos chanos* (milkfish) and tilapia. These ponds or "tambaks" are hand constructed by the natives often up to their waists in water and mud, the latter being compacted round the periphery of the enclosure to give side slopes of between 3:1 and 2:1. The sites chosen for construction are very critical from the point of view of pond management. The mudflat sites are chosen so that the material scooped out of the fish pond areas forms the embankments to a level above the high water spring tide level, leaving the fish pond bottom level at the low water neap tide mark. Thus the ponds can be emptied or filled at any time during the spring and neap tidal cycles.

However, there is a continual battle to maintain this type of construction against erosion, and vegetation of some description is required to hold the

Fig. 58   Brushwood mats on offshore face of embankment at Audenge mullet farm in the Bassin d'Arcachon.
(*Credit—P. H. Milne*)

mud together. Hickling (1962) reports that he has seen rows of trees planted on the pond sides, mangroves and tamarinds in Java and mulberry bushes in China, to try and shelter the ponds from the prevailing winds. It is essential that the pond waterline is protected to prevent erosion, and breakwaters of boards or lines of split bamboo have been used in Malaya, Java and China. Since 1939 several large fish ponds have been built in Israel on similar lines to the ponds in the Far East (Bull, 1963). However, these large ponds developed such a destructive swell in a fresh wind that the above techniques were ineffective, and stone pitching is now recommended for above and below the waterline to prevent erosion (Pruginin and Ben-Ari, 1959).

Mullet Culture in France

Mud embankments similar to the tambaks described above are also used to protect the fish ponds at Audenge in the Bassin d'Arcachon, which lies to the west of Bordeaux in France. These ponds are used for the farming of mullet as described by Arne (1938) and Dantec (1955). With a 4 m tidal range, and the embankments' exposure to winds, the toe of the mud embankment is protected by riprap, or stone pitching.

When I visited Audenge in 1965, I found that brushwood mats held in place by strands of fencing wire attached to wooden stakes were used for the protection of the embankment by dissipating the

wave forces, Fig. 58. However, this method requires regular maintenance, as the brushwood and stakes require replacement every three to four years. In some places a layer of asphalt had been used, as brushwood was becoming scarce and labour becoming more costly. (A discussion on the use of asphalt is given at the end of this chapter.) To stabilise the internal slopes of the embankments, bushes of tamarisk (an evergreen shrub, growing $1\frac{1}{2}$ to 3 metres high) had been planted to hold the soil together, Fig. 59, and these bushes acted as windbreaks to shelter the ponds in the hinterland. Turf was also used higher up the embankment to hold the soil together, and this method can be used to a certain extent to give a reasonable form of protection. Care must be taken in its installation, and regular maintenance is necessary to replace damaged turf, but given care and attention the Dutch have found turf to be perfectly satisfactory in not too exposed situations (Wiersma, 1960).

The success of the farming of mullet at Audenge lies partly in the overall design of the farm and its sluices. The ponds here were originally constructed to extract salt from sea water, but the salt merchants soon observed the development of the fish which also entered the pools. They eventually converted their brine pools at the beginning of the twentieth century into fish ponds, by building sluices to induce the wild stock from the Bassin d'Arcachon to swim into the ponds at the young fry stage. The 250 ponds at Audenge, covering an area of 150

Fig. 59   Bushes of tamarisk on top of the sea embankments at Audenge to shelter the mullet ponds in the hinterland.
(*Credit—P. H. Milne*)

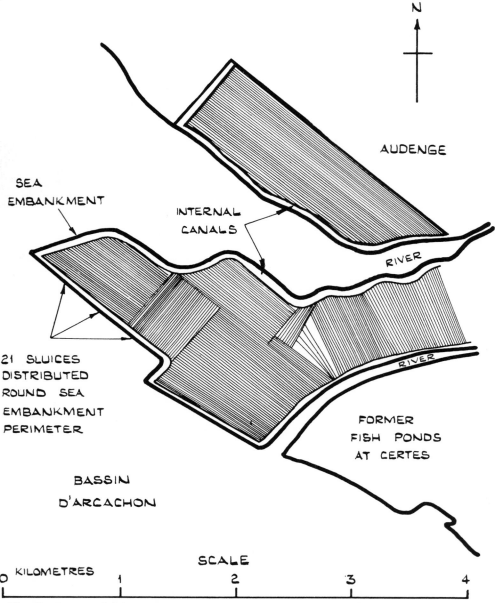

N

SEA
EMBANKMENT

INTERNAL
CANALS

AUDENGE

RIVER

21 SLUICES
DISTRIBUTED
ROUND SEA
EMBANKMENT
PERIMETER

RIVER

FORMER
FISH PONDS
AT CERTES

BASSIN
D'ARCACHON

SCALE

KILOMETRES
0   1   2   3   4

Fig. 60 Layout of fish ponds at Audenge in the Bassin d'Arcachon showing peripheral canal.

hectares generally lie in a N.W.–S.E. direction at right angles to the prevailing south westerly winds to prevent turbulence, as shown in Fig. 60. An internal canal connecting each of the fish ponds lies inside the sea embankment, round which are positioned 21 sluices for distribution of sea water or replenishment. In general the ponds are long and narrow, 200 m × 30 m (approximately half a hectare), with an average depth of 1 metre, Fig. 61.

There are, however, several ponds 2 metres deep into which the fish are herded during the winter months to counteract the colder temperatures.

The design of the sluice gates for a fish farm is one of the most important factors since it is the sluices that regulate the water movement and thus control pond management. The design and operation of the sluices at Audenge are the secret of their success, Fig. 62. These sluices are 1·10 metres wide

Fig. 61 Fish ponds at Audenge, average size 200 × 30 m (0·6 ha) with a depth of 1 m. (*Credit—P. H. Milne*)

and run the full width of the embankment. Two hours before the flood tide water level reaches the downstream end of the sluice (with the pond side fish net screen in position) the sluice gate is raised 7 cm to allow a trickle of water to run out of the ponds. The young mullet fry are attracted by this flow of water and swim up into the sluice. When sufficient young fry are in the sluice the seaward fish net screen is inserted to trap the fish, and once the tide level is greater than the pond level, the pond fish net screen is lifted to allow the young fry to swim into the pond. During this latter operation, the fish net sleeve is also inserted to prevent any fish from the ponds trying to swim into the sluice. The

sluices are operated in this manner to attract the white mullet in March, and the black and golden mullet from June to September (Arne, 1938). There are also two other frameworks which can be placed in the sluice for the catching of eels which are sold as a side line of the fish farm (Dantec, 1953). Normally when the sluices are used for the replenishment of sea water, a fine 7 mm mesh net attached to an oak wood frame is inserted upstream of the sluice gate.

To keep a depth of 1 metre in the ponds the water level must be maintained between the high water neap tide and the high water spring tide marks, and thus replenishment of sea water is only possible

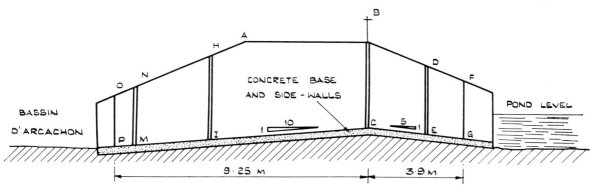

Fig. 62 Audenge Sea Sluice Gate details:
AB—roof of sluice carrying access road
BC—cast iron sluice gate
DE—position of fish net sleeve
FG—fish net screen—pond side

OP—fish net screen—seaward side
HI ⎱
NM ⎰ grooves for placing boards to catch eels as a side-line

(*Credit—from Dantec, 1953*)

six days per fortnight. Since the main river is diverted between embankments, and there is no catchment other than the ponds' surface, there is little reduction in salinity in the ponds, the average being 29·5‰ with a minimum of 23‰. One disadvantage of having interconnecting ponds, however, is that they cannot be individually drained for harvesting, and the mullet are fished out using a 40 mm net to ensure that only the mature fish are caught. The total annual production of these ponds averages 15 tonnes of mullet and 7·5 tonnes of eels.

Mullet Culture in Hawaii

The fish ponds of Hawaii have long been used for the combined cultivation of mullet, *Mugil cephalus*, and milkfish, *Chanos chanos*. However, production since the beginning of the twentieth century has dropped sharply from 220 tonnes in 1900 to 15·4 tonnes in 1960 (Iversen, 1968). Some of the early ponds are still in existence, and these were formed by closing off bays with dykes of coral and basalt, 1·5 m wide. This allows the water to percolate through the walls into the ponds which vary in depth from 0·1 m–1·0 m depending on the tide (Malone, 1969). Some of the mullet enter the ponds through the sluices and others are netted in nearby estuaries and placed in the ponds for growing and fattening.

The above passive farming method, however, relied on a constant supply of the fish for fattening, which was not always available. To overcome this problem the Oceanic Institute at Waimanalo has recently (1970) implemented a mullet breeding programme for *M. cephalus*. To control the water flow through the brood stock ponds, butyl-rubber liners have been used to form 9 m × 6 m × 0·9 m ponds on the shore (Shehadeh *et al.*, 1971). Only limited supplemental feeding of trout chow has been required since buoyant plastic sheets to develop diatom and algal growths for the mullets' diet have been used. These plastic sheets, anchored to the bottom, provide additional surface area for the attachment and growth of periphyton. This enables the more intensive stocking of fish, and provides a "free" supply of food for herbivorous and filter-feeding fish like the mullet (Shehadeh, 1970).

Salmon Culture in United States of America

As discussed in the previous chapter hatchery techniques are being developed for the raising of Pacific salmon, *Oncorhynchus* spp., in the United States. Experiments into the feasibility of rearing salmon in brackish water and salt water impound-

ments have, however, been carried out over the past decade.

Two such brackish water ponds have been constructed in Oregon State. The first pond at Lint Slough was built on intertidal mud flats near Waldport in 1962. The area enclosed covers 14 ha and a dam provides the control of the tidal waters. To control the salinity and prevent inundation by fresh water a large bypass canal was constructed to permit the diversion of Lint Creek round the new impoundment (DeWitt, 1969). The tidal variation in the pond gives mean depths of 2 m maximum and 1 m minimum. To permit the gradual change of salinity during rearing, several pumps are installed in addition to the tidal sluice gates to ensure adequate control. Both chinook and coho salmon, *O. tschawytscha* and *O. kisutch*, have been successfully raised in this impoundment. The second pond constructed at Cape Meares Lake, near Tillamock covers 37 ha. This pond has been used since 1968 for raising chinook salmon and has a very low salinity (DeWitt, 1969).

Eight salt water ponds have also been constructed in Washington State since 1958 for the raising of chinook and coho salmon. However, the early results of the rearing programme did not come up to the expectations of the Washington Department of Fisheries and in 1965 only four of these ponds were still in operation. These were at Crockett Lake (120 ha), Kingston Lagoon (7 ha), Titlow Lagoon (1·6 ha) and Whiteman's Cose (11 ha) (Phinney and Kral, 1965; Phinney, 1966). There were several factors causing this reduction including higher than expected construction and maintenance costs, fish and bird predation, high summer water temperatures and land ownership problems.

The potential for the economical cultivation of salmon in ponds and impoundments in the United States is still great but farmers must appreciate the biological, environmental, mechanical and managerial factors involved in running a successful enterprise (DeWitt, 1969).

Plaice Culture in Scotland

In Chapter 6 reference was made to the hatchery research work on plaice (*Pleuronectes platessa*) carried out at Port Erin in the Isle of Man for the White Fish Authority. By the summer of 1965 this research had reached a stage when the large scale cultivation of plaice larvae was possible, and a quarter of a million of these artificially hatched plaice, each 30 mm in length were used to stock a 2 hectare intertidal pond at Ardtoe, Argyll. This pond was designed for the White Fish Authority by

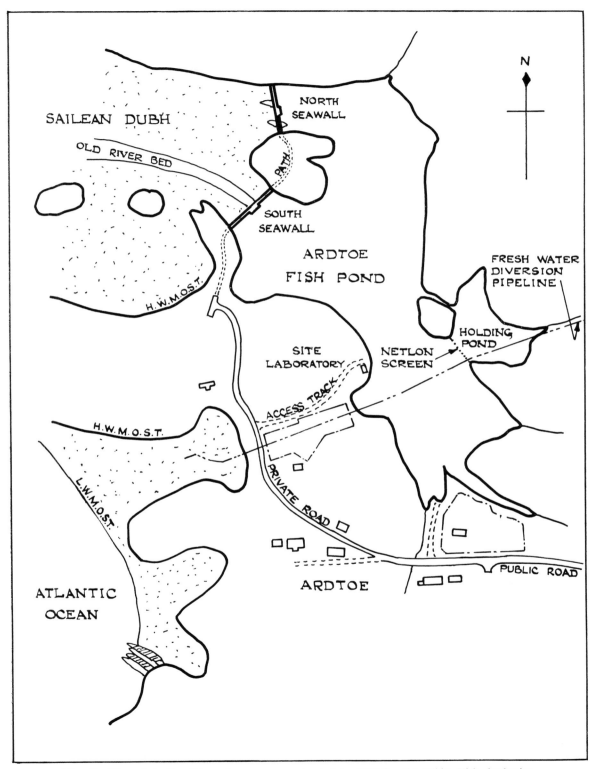

Fig. 63 Plan at Ardtoe fish enclosure, as constructed in 1965 for the White Fish Authority.

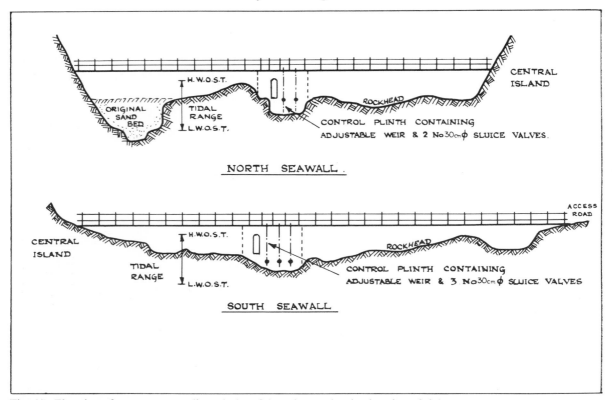

Fig. 64   Elevation of concrete sea walls at Ardtoe fish enclosure showing location of sluices.

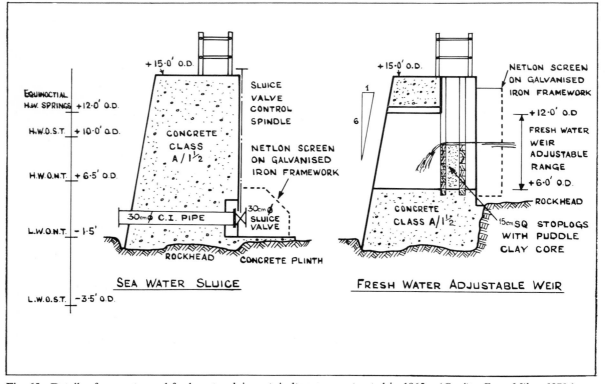

Fig. 65   Details of sea water and fresh water sluices at Ardtoe, as constructed in 1965.   (*Credit—From Milne, 1972c*)

Fig. 66 Ardtoe south seawall during construction. A monorail was used to place the concrete. (*Credit—P. H. Milne*)

members of the Department of Civil Engineering at the University of Strathclyde in Scotland (Allen and Milne, 1967).

The site chosen for Ardtoe lay in an area with a spring tide range of 4·3 m adjacent to north-west Atlantic sea water. The inlet at Ardtoe, called Sailean Dubh, Fig. 1, is capable of future developments, and of three possible locations, north, south and east, the latter was chosen for the 2 hectare enclosure. The main stream into Sailean Dubh flowed through this area so a small diversion pipe out to sea was provided to alleviate the fresh water inflow.

The intertidal area was enclosed using two concrete seawalls on rock foundations, either side of a central island (see Figs. 63, 64). Both seawalls had a 6:1 batter and featured low level sluices for the admittance of sea water and fresh water weirs to skim off the less dense rain water, as shown in Fig. 65. It is acknowledged that the stoplog method of forming an adjustable weir was not very sophisticated and was awkward for maintenance staff

to raise and lower, but it was the only conceivable method available in the six months between drawing board design and completion. Rectangular penstocks with adjustable weirs would have been preferable, but the minimum delivery date for completion of nine months was not acceptable. Subsequently the stoplogs have been replaced with a hand operated flap valve to ease operation and maintenance.

Concrete was chosen as the construction material since the rock in the vicinity was known to be difficult to blast and shape. To counteract the leaching out of the cement by the sea water, a richer mix was used (33 kg of cement per cu.m of concrete) than used normally on land. With an intertidal site like Ardtoe, it was possible to work in the dry at every low tide, thus dispensing with the need for expensive cofferdams. However, care must be taken to prevent the newly deposited concrete being scoured out by the rising tide, and the shuttering must be well anchored to prevent flotation, Fig. 66. The designer's programme of construction

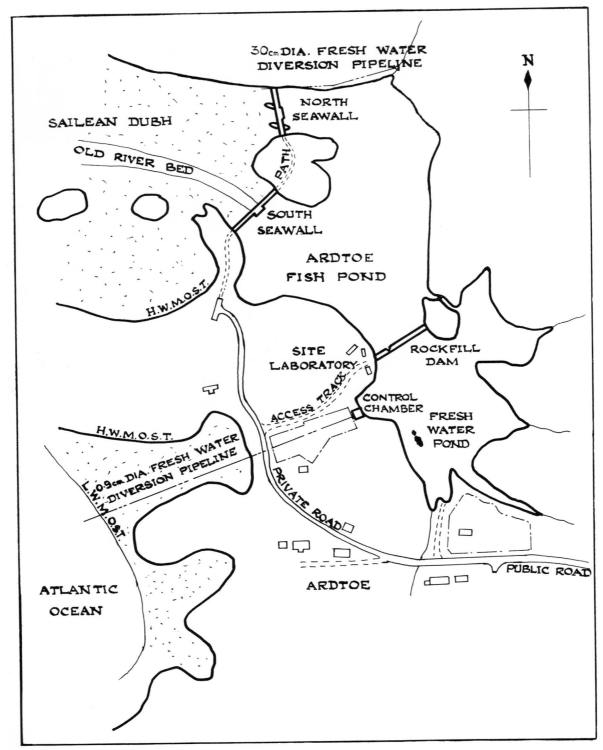

Fig. 67 Plan of Ardtoe fish enclosure as ammended in 1966 with a fresh water pond and diversion pipeline.
(Credit—From Allen and Milne, 1967)

must also be carefully adhered to, so that there is no danger of impounding water behind the seawall until the concrete is set hard, and such provision must be made for an adequate tidal flow round the structure.

The water level selected for the Ardtoe pond lay between high water neap tides and high water spring tides, only allowing for sea water replenishment six days per fortnight as at Audenge discussed earlier. In the autumn of 1965 difficulty was experienced in maintaining salinity as a north-westerly wind forced the fresh surface rain water into the south-east corner of the pond away from the sluices. It was thus impossible to drain off the brackish water from the south-east corner even during replenishment, due to the salinity wedge effect which was enhanced by the wind (Milne, 1971c), as discussed in Chapter 11.

Eventually continuous heavy rain during a period of neap tides, when no replenishment was possible, reduced the salinity throughout the whole pond. The answer to the prevention of inundation by fresh water is either to have a peripheral fresh water canal, or to divert all the main streams from the surrounding catchment away from the enclosure. The latter solution was adopted at Ardtoe, by the creation of a fresh water pond in the south-east corner with a pipeline diverting the water out to sea,

Figs. 12 and 67. A narrow neck of rock provided a suitable site for the erection of an internal barrier, but the seabed sediment was soft puddle clay. In this instance a rockfill dam with a puddle clay core was used, Figs. 68 and 69, and this dam was constructed in the drained pond in 1966 reducing the overall area to 1·2 hectares.

The White Fish Authority had hoped that by operating sluice gates for the exchange of sea water, a suitable marine environment could be maintained within the pond. It was also envisaged that the natural food available could be increased by the addition of fertilisers in a similar manner to earlier experiments by Gross *et al.* (1949, 1952) at Loch Sween in Scotland. In practice, this scheme proved too ambitious. Problems of environmental control and the inability to exercise proper husbandry over the fish and indeed the inability to harvest the fish satisfactorily, has led to a move towards high density culture in smaller permeable enclosures and cages (Richardson, 1971), as discussed later in Chapters 8 and 10.

Sea Bream Culture in France

With increasing interest in sea farming in France, a private company, Compagnie des Salines du Midi, has established in Sète, an experimental unit to carry out feasibility studies into the farming of

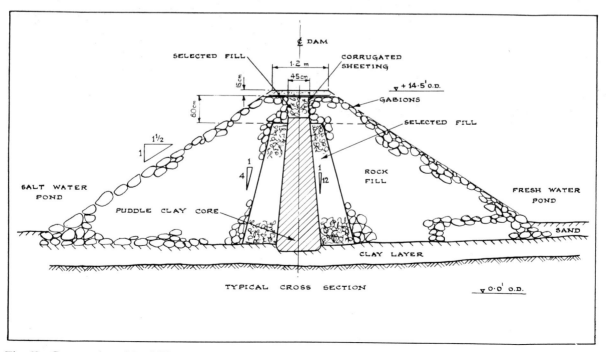

Fig. 68   Cross-section of Rockfill Dam at Ardtoe, showing use of puddle clay core as impervious barrier.

Fig. 69   Rockfill Dam at Ardtoe with concrete walkway and small sluice gate.   (*Credit—P. H. Milne*)

sea bream, *Chrysophrys aurata* (Serene, 1969). This company was established in 1969 and are using disused salt ponds for their sea enclosures.

At present sea bream fry are caught from the Bassin de Thau, but due to objections from local fishermen, the company are now developing their own hatchery brood stock.

After capture, the wild fry are held for 7–8 days in 4 m circular tanks, 1 m deep. These tanks are of a very simple construction using three plywood sheets curved to the tank diameter with the bottom edge sunk into the sandy soil. The plywood sheets are strengthened with fencing wire and strainers to prevent buckling, and the tank lined with polythene sheeting. The fry are starved for the first three days and then an automatic feeder is used to accustom the fish to the pelleted food. After this weaning period the fry are moved to the stocking enclosures.

These disused salt ponds, average size, 0·4 ha have been adapted for sea bream farming. As the main ponds were only 10–20 cm deep a 4 m wide moat has been dug 2 m deep round the periphery. The earth removed has been used to provide stable bunds which have been staked and planked for access vehicles. The fish are said to bask on the shallow platform in the centre in the summer and retire to the deeper water in the winter. Four automatic feeding dispensers are provided around the periphery.

The target production of sea bream from Sète is 150 tonnes per annum with an average of 1 kg per square metre, the ponds being stocked in June and harvested the following April on a one year cycle basis.

Intertidal Barrier Construction

Many fish and shellfish farming enclosures have been constructed, especially in the Far East in the intertidal zone of the seashore for the pond culture of brackish water fish. The design and construction of these ponds, generally constructed from mud and clay, was discussed earlier during the description of intertidal ponds.

Materials other than mud and clay, which present a very unstable surface for general engineering plant construction (unlike the hand methods used in the Far East), have been considered for the building of intertidal enclosure barriers.

As there are many large sandy bays and estuaries surrounding the coastline, the possibility of constructing enclosures using sand as the basic retaining embankment material has been considered. The handling and compaction of sand is readily carried out, the only problem is its porous nature. To prevent seepage through the sand there are two methods available. The first is to provide an overall blanket such as polythene sheeting or reinforced concrete, and second to provide a cut-off wall in each embankment, to lengthen the seepage path by either steel sheet-piling or cement grouting. The exposed offshore face of the embankment will certainly need protection, and depending on the size

of the enclosures, the inside faces at the waterline may need protection.

Two interesting papers by Burgers (1926) and by Rasmussen and Lauritzen (1953) highlight the problem of seepage from irrigation canals, and how it is possible to minimise these losses. The provision of sheet-piling cut-off walls to give a seepage seal can be carried out simply as reported by Bayard (1964) and by Pugnet and Capitaine (1964). The art of providing grout cut-off walls has now been well developed (Symposium, 1964), and could be readily applied.

Especially if small ponds are desired, the concept of polythene sheeting merits attention. The use of thin flexible impermeable membranes to prevent the seepage of water has been developed mainly in the United States (Staff, 1967).

The materials used are plasticised polyvinyl chloride, polyethylene and butyl rubber. The sand itself will generally support high hydrostatic loads, and the sheeting employed as a membrane will

Fig. 70   Rainbow trout earth ponds using a PVC-vinyl liner at Kenmure Fisheries, Loch Ken in Scotland.

(*Credit—P. H. Milne*)

follow earth movements and settlements while retaining its impermeability. Vinyl liners have been used for: salmon rearing ponds in Washington (Staff, 1967), Figs. 43–46; rainbow trout ponds in Scotland, Fig. 70; water storage ponds in Arizona; a waste disposal basin in New Jersey (Anon., 1966a,d); an artificial reservoir in Ontario (Anon., 1962); shrimp and pompano ponds in Florida, Figs. 32–34 (Tabb *et al.*, 1969); and for prawn ponds in England, Fig. 36, described in Chapter 6. The normal installation procedure is to grade the embankments to a side slope of 3:1, and to cover the liner with topsoil, sand in the case of fish ponds. The liners are easily joined *in situ*, so that there is no restriction on area. Connections to pipes and other structures are quite easily made with adhesives and pieces of the sheeting. This method of construction would be suitable for any area where the material in the intertidal zone was capable of being contoured by earth moving machinery.

The offshore face of the main embankment, exposed to the rigours of the tides, wind and wave movement will require protection, and this subject has received considerable attention from Bruun (1953, 1964), and is detailed in the development of

the construction of seawalls along the coastline, Fig. 71, by Thorn (1960). However, the development of synthetic filters (Agerschou, 1966; Anon., 1966), and asphalt (Lint, 1964; Asbeck, 1964; Asphalt Institute, 1961, 1962), to form protective layers on embankments for coastal protection have been developed in the U.S.A. and the Netherlands in the last decade. These methods give a simple yet reliable form of construction for protection, and have already been used at the fish ponds at Audenge, described earlier in this chapter.

The use of traditional methods for the construction of dams and seawalls of stone or concrete can be readily applied to the construction of sea farming enclosures. This type of construction is generally confined to sites where rockhead is known on the shore to give a solid foundation. The choice of material is dependent on the availability of stone of a suitable size, and the accessibility of the site for the transportation of cement and materials (Slichter, 1967). If a solid rock foundation is assured over the length of the structure, and cement and concrete aggregate are readily available it may well be possible to construct the whole of the seawall in the dry (Duhoux, 1964). An intriguing combina-

Fig. 71 Rockfill embankment with placed stone offshore face and concrete parapet at Longannet, Firth of Forth in Scotland. (*Credit—P. H. Milne*)

Fig. 72 Intertidal lobster enclosure at Crookhaven in Ireland. The water level is maintained at high water neap tide with a concrete wall, with mesh netting thereafter, allowing daily tidal flushing. (*Credit—J. H. Allen*)

tion of concrete up to high water neap tide level and mesh thereafter has been built for temporary lobster storage at Crookhaven in Ireland, Fig. 72.

The fish enclosure at Ardtoe, Scotland, was designed in 1965, with concrete seawalls, Figs. 65, 66, as the rock was so hard (moine schist) that blasting and shaping would have been very difficult and time consuming, and road access for transportation of cement was available. Sometimes savings can be made on the quantities of materials if prestressing cables are used to anchor the seawall to the foundation (Bellier, 1966; Cambefort, 1966; Chaudesaigues, 1966).

If, however, excavation to rockhead is not practical, the simplest form of construction is to utilise rockfill with an impermeable core. This core can be formed by sheet-piling or grouting or adopting a puddle clay core as used at Ardtoe in 1966 to form a fresh water pond, at the south-eastern extremity of the enclosure, Figs. 68 and 69.

The dumping of rockfill can either be placed or tipped by lorries and bulldozed into position (Blanchet, 1946a, b).

The rock used for seawalls requires to be specially quarried and graded for the specific purpose in hand. Rock for rockfill should be hard, dense and durable, and should be able to resist long exposure to weathering. The individual pieces must be of sufficient weight to resist displacement by wave action. The shape of the individual stones or rock fragments influences the ability of the riprap to resist displacements by wave action (Hydraulics Research, 1964). Angular fragments of quarried rock tend to interlock and resist displacement better than do boulders and rounded cobbles.

Quite often from the analysis of the expected waves it is found that there is no rock of a suitable size in the area, to protect the core of the seawall, and recourse to man-made components, such as concrete blocks (Nagai, 1962; Anon., 1966b; Jachowski, 1964; Hall, 1968), tetrapods (Danel *et al.*, 1960; Danel and Greslou, 1962; Kjelstrup, 1963), tribars (Danel *et al.*, 1960, 1962), stabits (Singh, 1968) or akmon (Paape and Walther, 1962), units is required. Recent research by Olivier (1967) and Izbash *et al.* (1970) into the development of rapid methods for the design and construction of rockfill dams has established methods for carrying out construction allowing for overtopping.

Chapter 8
Sublittoral Farming Techniques

If it is not essential to control the environment of the farmed species, the water movement created by tidal currents can be used to advantage in the sublittoral zone. In general the technique used for sea farming in the sublittoral zone is the enclosure of natural bays using either embankments with mesh sluices or net barriers. The former method is used for oyster and yellow-tail culture and the latter method for yellow-tail, rainbow trout and pompano culture.

The design of a sublittoral enclosure for sea farming is determined by its location and may take several forms (Milne, 1970e):

(i) completely isolated enclosures surrounded by a net structure in the middle of a bay with no foreshore;
(ii) a shore enclosure with a portion of the foreshore extending into deep water surrounded by a net structure; and
(iii) a sea bay or loch enclosure with an embankment or net structure only at the entrance.

Examples of each type of enclosure are described for various species.

Due to an increasing interest in the use of sublittoral enclosures for fish farming, a detailed account is given of research carried out in Scotland into various construction techniques suitable for the sublittoral zone, depending on location and exposure. In some locations a boat lock may have to be provided for the passage of small craft, and their construction is also discussed.

The only marine organism not using an impoundment technique in the sublittoral zone is the green mussel of the Philippines which requires to be immersed at all states of the tide.

(I) MOLLUSCS

Oyster Culture in United States of America

Hatchery techniques for the production of oyster spat were described in Chapter 6. One company,

The Ocean Pond Corporation, Fisheries Island, however, uses a sublittoral pond for the controlled production of oyster spat (Matthiessen, 1969). A natural spawning population of American oysters, *Crassostrea virginica*, is kept in a 9·3 ha brackish water pond. Normally the pond is open to the ocean, but just prior to spawning in early summer, the sluices are closed to prevent the escape of larvae and to accelerate the rise in water temperature. When sufficient quantities of the oyster larvae are present, 2·4 m long shell strings, consisting of scallop shells threaded on galvanised wire, are suspended in the water from wood and styrofoam rafts. After the spat has set the pond is reopened to the sea, and the young oysters held in the pond until the following spring, when they are transferred to private oyster beds in Long Island Sound, as described in the next chapter.

This firm have been in operation since 1962 and by 1969, a total of 60,000 shell strings were suspended in the pond with an annual production averaging in excess of 50 million oysters for planting.

Oyster Culture in Russia

At the beginning of the century the Black Sea had a large natural population of oysters, with an annual harvest of 10 million oysters. However, by 1912 oyster fishing had ceased due to complete absence of control. Recent experiments on the artificial cultivation and breeding of oysters in salt water ponds has produced superior oysters to those still existing naturally (Anon., 1970f).

Oyster cultivation by the Azov-Black Sea Institute of Fisheries has centred at Yagorlitsk Bay on the west coast of the Black Sea. After carrying out successful trials in an experimental pool, a large oyster breeding farm with three pools and a total capacity of one and a half million oysters has been constructed (Ziatsev, 1970).

The method adopted is to put young oysters in cages, which are suspended from wires stretched between posts driven into the seabed. Growth rates to 60 mm in diameter after one year have been achieved with 70 mm in two years. Although Black Sea oysters are smaller than those of other countries, they are said to excel in quality.

Mussel Culture in the Philippines

The cultivation of mussels in the Manila Bay region of the Philippines has developed as an offshoot of oyster culture, discussed in the previous chapter. When settling surfaces were prepared for oysters they were often fouled up with the green mussel, *Mytilus smaragdinus*. Although wild mussels had often been gathered for food, no previous thought had been given to their culture, until it was discovered that they could be cultivated on vertical poles driven into the seabed in a similar manner to that developed for oyster culture (Ryther and Bardach, 1968).

Mussel culture, however, requires sublittoral conditions since the green mussels do not withstand intertidal exposure to the air. Thus mussel culture can be carried out adjacent to, but offshore of oyster culture areas. The technique adopted is to drive dried bamboo poles, 5–10 cm in diameter into the seabed in water depths ranging from 2–8 m at low water. The poles are used as the spat collectors in the same way as the oyster culture, and the mussels set on the bamboo poles in the spring. The green mussels grow much more rapidly than the oysters, and reach a market size of 5–7·5 cm in 4–8 months. They must therefore be harvested in the first summer, or the clumps, up to 0·3 m in diameter, are liable to become detached due to their weight, with their subsequent loss in the soft sediment.

As this is a new industry very little data is available on production, but Ryther and Bardach (1968) reported that in 1966 there were some 60 fishermen cultivating 8 ha of the seabed, with a production total of 250 tonnes per hectare. The potential yield from such a location could thus be increased by using hanging techniques as discussed in Chapter 10.

(II) CRUSTACEANS

Shrimp Culture in the United States

As mentioned in Chapter 6, Marifarms Inc., established a large hatchery in Panama City in 1968 for the raising of 400 million shrimps a year for stocking their 1,000 ha sublittoral enclosure in the West Bay area of the Gulf of Mexico (Webber, 1970).

The area has been enclosed by a net barrier of mesh netting suspended from cables strung between piles jetted down into the seabed, similar to the Japanese net enclosures for yellow-tail. Since the 10-year lease stipulates that public access is maintained, a 6 m section of the net is designed as a boat lock, and is manned round the clock. The gate is weighted to sink when released and is raised by means of a hand cranked winch. By intensive restocking, feeding, and protecting the area from natural predators, Marifarm expect to produce 1,125 kg of shrimp per hectare per annum, which compares very favourably with intensive pond culture (Anon., 1970c, 1971n).

Unfortunately the 1970 harvest from the pond was not very good since although stocked with 175 million post-larval shrimp, the harvest was only 7·7 tonnes. Two reasons given for this were that delays in obtaining the lease resulted in premature stocking of the enclosure before all the predators were removed and also trouble with the mesh netting. The initial nets used were not resistant to marine fouling and became overloaded with fouling beyond Marifarm's ability to keep it clean. The net has now been completely replaced by a new, creosote-impregnated netting.

The enclosure technique at West Bay is to release the post-larval shrimp into small areas enclosed by fine mesh netting. As the shrimps grow these nets are removed, releasing them into progressively larger areas enclosed by nets with progressively larger meshes, up to 16 mm. To avoid any loss of shrimp, or trouble with fouling, the nets are patrolled regularly by skin divers. Any fouling is cleaned off using high-pressure jets underwater, although with the new creosoted twine this is now less of a problem.

The harvest from the company's enclosures in 1971 reached 226 tonnes which were collected in the normal way using Laffitte trawlers.

(III) FISH

Yellow-tail Culture in Japan

The marine cultivation of yellow-tail, *Seriola quinqueradiata*, accounted for 98·6% of the total production of cultivated marine fishes in Japan in 1967 and amounted to 27,139 tonnes (Harada, 1970).

Three types of enclosures are used in this cultivation:
  (i) pond enclosures in the sublittoral zone formed by embankments with sluices for water circulation;

(ii) netting enclosures in the sublittoral zone using net barriers to partition off areas of the seabed; and

(iii) floating net cages moored in sheltered bays.

The first two types of enclosures will be discussed here and the floating net cages, which are used to a greater extent, due to their simplicity, are discussed in Chapter 10.

Adoike The majority of Japanese fish farms are concentrated along the south-west coast and the Inland Sea region due to the warmer water temperatures. Here several bays and channels have been

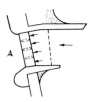

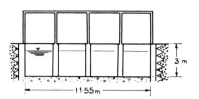
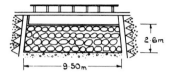

Fig. 75 Elevations and plans of sluice gates and screens at Adoike yellow-tail farm in Japan. The location of gates A and B is shown in Fig. 73.
(Credit—From Tamura and Yamada, 1963)

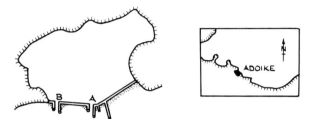

Fig. 73 Plan of 27 ha yellow-tail enclosure at Adoike near Takamatsu on the Inland Sea in Japan. The embankment has two seawater sluice gates at A and B.
(Credit—From Tamura and Yamada, 1963)

enclosed, using embankments of stone, earth or concrete depending on the availability of materials, and the exposure to storms. The largest of these pond enclosures is at Adoike near Takamatsu on the Inland Sea, and covers an area of 27 ha, Figs. 73 and 74. The average depth of this pond is 8 m and it is continually open to the sea via two

sluice gates with screens, Figs. 75, 76. This pond has been established for over 50 years, and is stocked each year with a quarter of a million yellow-tail fry, 2·5 cm long and 3 g in weight. The fry are fed trash fish up to their marketable weight of about 1 kg. Fry stocked in enclosures from May to July grow to 200–700 g by the end of August, 600–1,600 g by the end of October, and 700–2,000 g by the end of December (Harada, 1965). It is interesting to note that the price of cultivated fish raised in enclosures is higher than those caught in the sea. The main reason for this is that the taste and flesh quality of the farmed fish is better than the wild stock (Ryther and Bardach, 1968).

Fig. 74 Adoike yellow-tail farm in Japan showing embankment and sea water sluices. *(Credit—J. D. M. Gordon)*

Fig. 76   Close-up photo of sluice gate A at Adoike showing mesh screens and lifting gantry.   (*Credit—J. D. M. Gordon*)

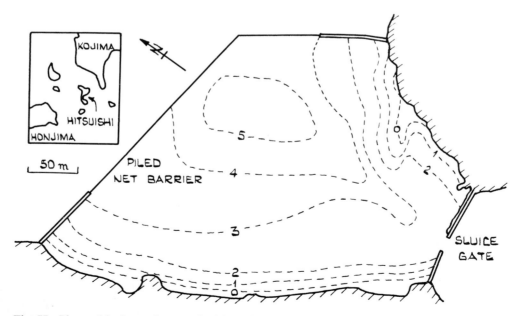

Fig. 77   Plan and bathymetric map of 7·2 ha yellow-tail enclosure at Hitsuishi near Tamano in Japan, showing 100 m embankment at southern end and 350 m piled net barrier at northern end.

(*Credit—From Inoue* et al., *1966*)

With the high cost of construction of embankments in the sea, the enclosure of areas using net barriers has proved popular due to their lower building costs. Numerous net enclosures now exist in Japan, the main ones being at Hitsuishi, Ieshima, Matsumigauru, Megishima and Tanoura. Owing to an increasing interest throughout the world in the use of net barriers for fish farming, each of these enclosures will be discussed separately.

Hitsuishi At Hitsuishi, near Tamano, the channel between a large and small island, Fig. 77, was sealed off to provide a 7·2 ha enclosure for yellow-tail. The tidal range in the area is 2·4 m with a maximum depth in the centre of the bay of 5 m. An embankment, 100 m long with sluices was chosen to protect the farm against storm damage at the southern end and at the more sheltered northern end a net barrier, 350 m long, is stretched between the two islands (Inoue *et al.*, 1966). This net barrier is hung from a framework of 60 cm diameter concrete piles, 12 m long, driven into the seabed 7·5–8 m apart, 42 piles being used in the 350 m length (Milne, 1970b,e). A vinyl covered net, with a 40 mm mesh is hung from 19 mm steel cables attached to the top of the piles which are 1·5 m above high water, Fig. 78. To restrain the lateral movement of these piles some are anchored fore and aft, using 19 mm steel cables, to large concrete anchor blocks, 100 cm × 100 cm × 120 cm in size,

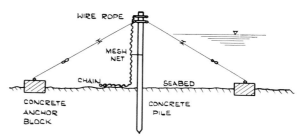

Fig. 79 Cross-section of piled net barrier at Hitsuishi showing restraining anchor blocks and weighted mesh net. Location of net barrier shown on Fig. 77.

(Credit—From Inoue et al., 1966)

Fig. 79, and the nets are attached at the low water mark, and weighted at the bottom by heavy chains to maintain a seabed seal. As in the majority of net enclosures, a boat lock is included in the net barrier for access. Unfortunately due to the large tidal range, 2·4 m, and the shallowness of the basin at low water the variation in volume from $22·3 \times 10^4$ m³ at high water to $8·7 \times 10^4$ m³, at low water, restricts the stocking density of yellow-tail to ensure sufficient dissolved oxygen and circulation for the fish.

Ieshima Ieshima fish farm, in the Hyogo Prefecture consists of an open bay, 91·7 ha in area sealed by a 400 m long length of net barrier, Fig. 80, with depths in the centre of the barrier of 15 m,

Fig. 78   Close-up view of net barrier at Hitsuishi showing net hung from steel cables strung between concrete piles.

(Credit—J. H. Allen)

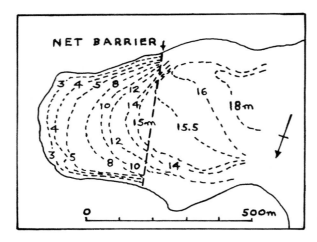

NET BARRIER

3 4 5 8 12 16
10 14 18m
15m 15.5
4 12
3 5 8 10 14

0 500m

Fig. 80 Plan of 91·7 ha yellow-tail fish enclosure at Ieshima in Japan, with 400 m long net barrier.
(*Credit—From Sugimoto* et al., *1966*)

the major part of the fish farm occupying depths of 4–15 m (Sugimoto *et al.*, 1966). This farm normally carries a stock of 600,000 yellow-tail, with a tidal variation of 1·2 m, which stimulates the circulation and water movement. Measurements on the water current speeds within the bay indicate a reduction to 46% of the current speed outside due to the resistance of the net barrier, and the possible reduction in water flow must be considered in the design of net enclosures.

Matsumigauru The largest area to be enclosed by a net barrier lies at Matsumigauru in the northwest corner of Hamana Lake. This enclosure covers 120 ha and is almost landlocked with a narrow entrance 300 m wide, Fig. 81 (Allen, 1968). The operators describe this enclosure as a ranch rather than a farm since they herd not only yellowtail but also puffer fish and sea bream in a mixed species enclosure. The net barrier is not placed straight across the entrance but set back as shown on Fig. 81, since the operators believe this helps to increase the circulation since there is only a 20 cm tidal range.

The deepest part of the net barrier is in 9 m of water, and steel piles, 45 cm in diameter at 20 m centres, were driven down to rock, leaving 2 m above mean sea level. The nets are suspended from an "I"-beam welded to the top of the piles, which also supports a working platform (Milne, 1970e). Lateral support for the net barrier is provided at every third pile with cables tied back to anchor blocks, as described for Hitsuishi earlier.

When the farm was constructed in 1965 two polythene mesh nets were used on the central part of the net barrier and a single vinyl-covered wire net used at the sides; all nets were 15 mm mesh. The operators thought that only the centre nets would foul, and chose polythene mesh as it was easier to clean. In fact all nets fouled heavily with ascidians in the first year. Allen (1968) reports that in September 1965 a typhoon with winds of 25 m/s crossed Hamana Lake. These winds raised the water level in Matsumigauru, and when the wind shifted with a drop in the water level of Hamana Lake the fouled nets could not withstand the fast flow of water and were carried away. A subsequent enquiry concluded that the nets on either side should be cleaned regularly as well as the centre, and that double rather than single nets should be used in all cases to facilitate cleaning. In addition a trash net was also placed on the offshore side of the centre portion to cater for floating debris liable to damage the fine mesh retention nets.

Megishima The fish enclosures at Megishima, near Takamatsu are constructed alongside a harbour mole with the shore on one side. The area is divided into one large and two smaller enclosures covering an area of 5·6 ha connected by boat locks (Sugimoto *et al.*, 1966), Fig. 82. The majority of these enclosures are shallow with depths of 2–5 m. Three rubble mound breakwaters were erected offshore to protect the net barrier from wave action.

The net barrier at Megishima is constructed from 40 cm diameter steel piles driven at 15 m centres. Here, two nets of galvanised wire mesh are used, the outer one with a 5 cm diamond mesh and the inner one with a 2·5 cm square mesh. The nets are suspended from 35 mm diameter steel bars placed at 2 m intervals from the top of the piles 2 m above mean sea level, Fig. 14. The nets were anchored at the seabed by dumping rubble on top of a short return section, as shown on Fig. 82. To improve lateral support some of the piles have cables attached to anchor blocks offshore. To assist in the cleaning of the net, guide ropes are fixed from a bar at the top of each pile to enable a boat to be pulled along the netting. Out of a total staff of 25 on this farm, seven were full time divers employed on cleaning the 1,370 m length of netting, Fig. 14.

Tanoura The Tanoura fish farm located on the southern coast of Syozu-sima in Kagawa Prefecture has a long bay mouth, 560 m wide, and is enclosed with double nets of 20 mm and 10 mm,

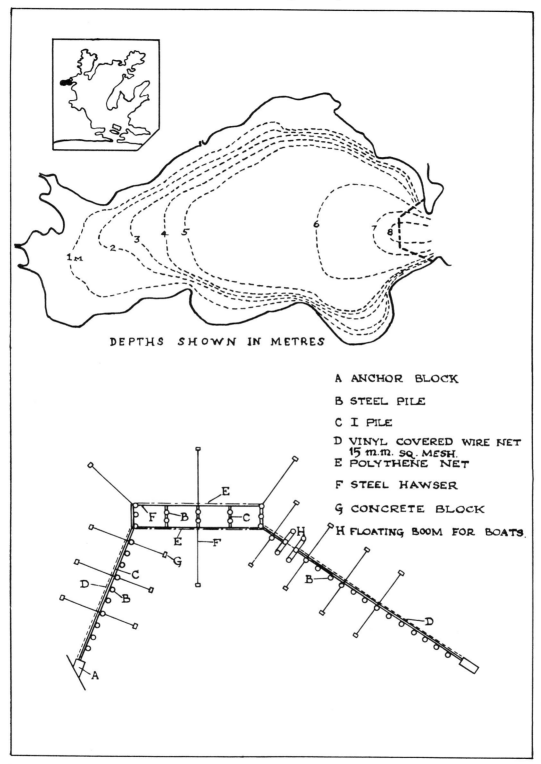

DEPTHS SHOWN IN METRES

A ANCHOR BLOCK

B STEEL PILE

C I PILE

D VINYL COVERED WIRE NET
15 m.m. SQ. MESH.
E POLYTHENE NET

F STEEL HAWSER

G CONCRETE BLOCK

H FLOATING BOOM FOR BOATS.

Fig. 81 Plan and location of netting barrier at Matsumigauru Fish Farm in Hamana Lake, Japan.
(*Credit—From Allen, 1968*)

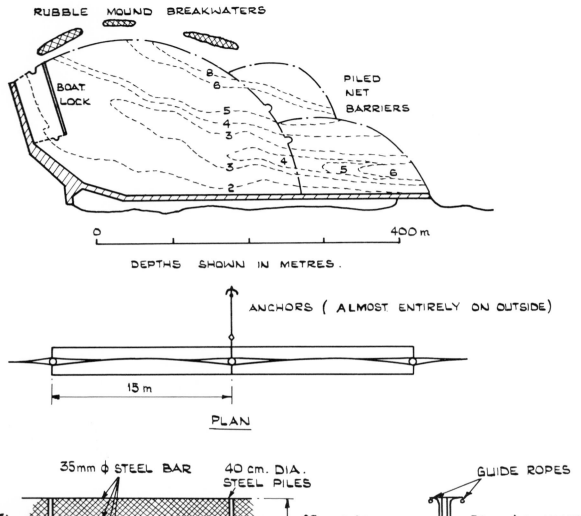

RUBBLE MOUND BREAKWATERS

BOAT LOCK

PILED NET BARRIERS

8
6
5
4
3
3
2
4
5
6

0   400 m

DEPTHS SHOWN IN METRES.

ANCHORS (ALMOST ENTIRELY ON OUTSIDE)

15 m

PLAN

35mm φ STEEL BAR   40 cm. DIA. STEEL PILES

GUIDE ROPES

M.S.L.

8 m

25mm □ GALVANISED WIRE MESH INSIDE

50mm φ GALVANISED WIRE MESH

RUBBLE PILED ON TOP OF NET

ELEVATION   SECTION

Fig. 82 Plan of yellow-tail fish farm at Megishima, near Takamatsu showing three net enclosures, with details of net fixing arrangements. (*Credit—From Allen, 1968*)

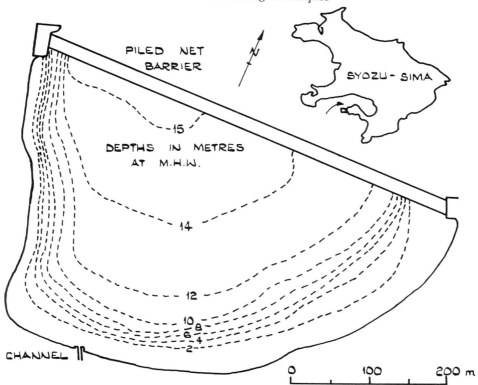

Fig. 83 Plan and location of yellow-tail fish farm at Tanoura on southern coast of Syozu-sima, showing 560 m net barrier across bay. (*Credit—From Inoue* et al., *1970*)

Fig. 84 Rainbow trout fish farm at Strom Loch in Shetland. Bag nets are hung from scaffolding poles at catwalk height.
(*Credit—P. H. Milne*)

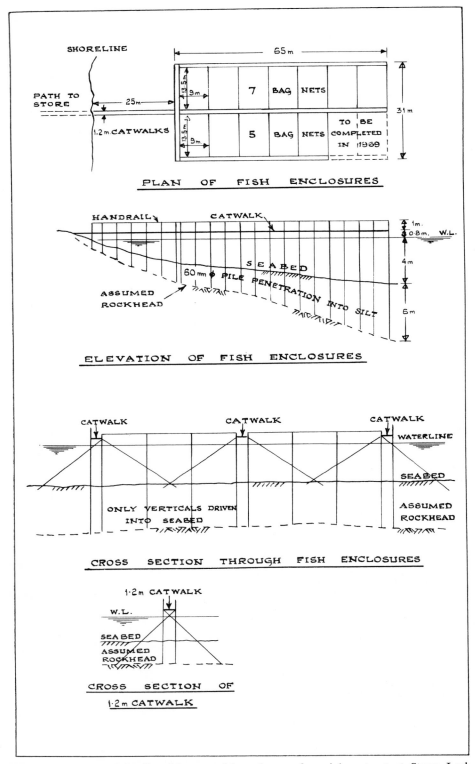

Fig. 85 Layout and details of bag net fish enclosures for rainbow trout at Strom Loch in Shetland. (*Credit—From Milne, 1972c*)

in two rows, hung from 50 cm piles at 10 m centres. This is an extremely deep enclosure with depths of 15 m, as shown in Fig. 83, covering an area of 16·4 ha (Inoue *et al.*, 1970). On the estimated water exchange given by the 1·5 m spring tidal range Inoue *et al.* reckon that the optimum stocking density was 40 × 10⁴ yellow-tail when weighing 400 g. The figures for stocking and harvesting in 1965 were 46 × 10⁴ and 42 × 10⁴ fish showing a very reasonable comparison. The total production of a fish farm thus depends on the water exchange rate at neap tides and this should always be considered in estimating production totals for a certain area.

Rainbow Trout Culture in Scotland

As mentioned in Chapter 6, Howietown and Northern Fisheries Limited chose a sublittoral technique for the growing and fattening of their rainbow trout fry, *Salmo gairdnerii*, in salt water. The site selected lay on the west shore of sheltered Strom Loch at the head of Stromness Voe in Shetland. By utilising mesh bag nets suspended from a fixed framework adjacent to the shore of the sea loch, Figs. 84, 85, access is not dependent on the weather, and the fish may thus be fed at all times. The reason Strom Loch was selected was its proximity to Lerwick, 18 km away, with a fishing port and fish processing factory, and also the small tidal range of the loch. Although the spring tidal

range in the Shetlands is 1·5 m, due to restrictions at the entrance to Stromness Voe and at Strom Loch, the tidal range is reduced to 0·3 m.

The method employed at Strom Loch resulted from research carried out by members of the Department of Civil Engineering at Strathclyde University (Milne, 1970b,e). This research investigated suitable methods of constructing frameworks for fish farming in sublittoral conditions, and recommended the use of individually driven galvanised scaffolding poles connected by horizontal members placed at selected heights. At Strom Loch the water depth varied from 3 to 5 m and the framework was designed for the suspension of bag nets, 9 m × 13·5 m × 3 m of 1 cm Courlene or nylon mesh fabric suspended from the scaffolding poles at catwalk height, Figs. 85, 86 (Milne, 1970b,e).

The rainbow trout fingerlings to stock these enclosures, 10,000 to a bag net, came from the hatchery in Stirling, as detailed earlier in Chapter 6. The food for feeding the fish comes from the

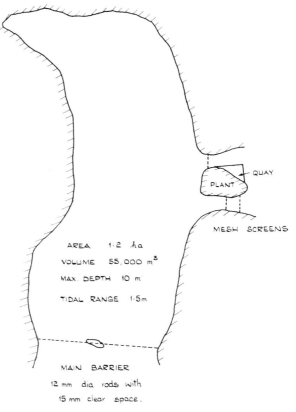

Fig. 86   Catwalk between rainbow trout bag nets hung from scaffolding poles at Strom Loch in Shetland.
(*Credit—P. H. Milne*)

Fig. 87   Plan of 1·2 ha sublittoral enclosure at Flogøykjölpo in Norway for salmon farming. (*Credit—by courtesy of A/S Mowi*)

Lerwick fish processing factory, and in 1968 over 1½ tonnes of trash fish were fed to the rainbow trout each day. The production in 1968 from these 14 bag nets was 10 tonnes, in 1969, 11 tonnes, and in 1970, 16 tonnes (Milne, 1970c).

It was hoped at the time of construction that the production from this farm would be 55 tonnes per annum. Unfortunately the hydrography of the area was not checked before construction, and trouble has since been experienced due to lack of water movement and circulation. A study of the hydrography (Milne, 1971c) has shown that the sea loch chosen with such a small tidal range, 0·3 m, was similar to many of the stagnant fiords in Norway. At present (December 1971) the farm has ceased production, but it is hoped that by the provision of mechanical aeration, as adopted in Norway, the problem of water circulation will be overcome.

Salmon Culture in Norway

Two Norwegian concerns have recently commenced the farming of Atlantic salmon, *Salmo salar*, in sea water enclosures.

The first sublittoral enclosures were built by A/S Mowi at Flogøykjölpo and Veløykjölpo, near Mövik on the island of Sotra, west of Bergen. Another set of smaller enclosures have been constructed by Eros Laks A/S at Bjordal in Sognefjord to the north of Bergen. These areas with tidal ranges of 1·5–2 m are very suitable for this type of enclosure.

Flogøykjölpo This 1·2 ha enclosure, Fig. 87, was completed in 1969 and has a production capacity of 150,000 kilos. The area is enclosed by two sets of barriers at the mouth and across subsidiary side channels. The main barrier, Fig. 88, is formed by concrete pedestals 60 cm wide by 100 cm, constructed at 10 m intervals across the bay, in 4 m of water. "I"-beams with a 16 cm flange span between the pedestals and are placed at 50 cm vertical intervals. The fish seal is provided by 12 mm dia. steel rods, with 15 mm clear space, inserted vertically through holes in the centre wedge of the "I"-beam, thus allowing for variations in the seabed level between the pedestals. The subsidiary side channels are sealed by mesh screens inserted in

Fig. 88 Flogøykjölpo salmon netting enclosure in Norway. Main barrier in distance, subsidiary mesh screens in foreground, with plant, and pumps to stimulate circulation. (*Credit—by courtesy of A/S Mowi*)

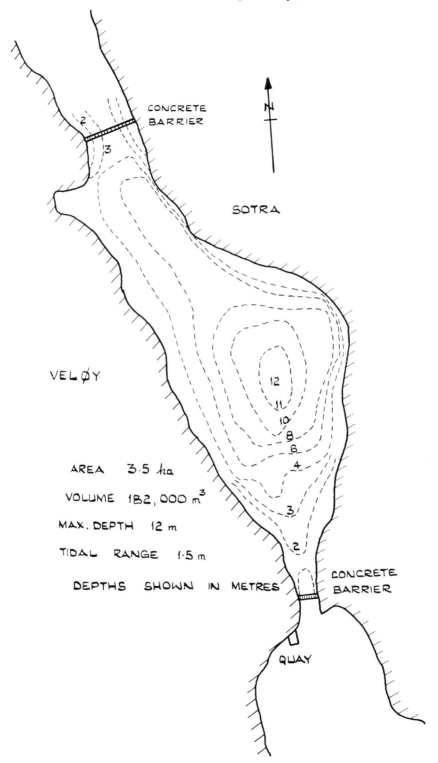

CONCRETE
BARRIER

N

SOTRA

VELØY

12
11
10
8
6
4

3

2

CONCRETE
BARRIER

AREA   3·5 *ha*

VOLUME   182,000 m³

MAX. DEPTH   12 m

TIDAL RANGE   1·5 m

DEPTHS SHOWN IN METRES

QUAY

Fig. 89   Plan of 3·5 ha sublittoral enclosure at Veløykjölpo in Norway for salmon farming.
(*Credit—by courtesy of A/S Mowi*)

guideways, shown at the bottom in Fig. 88. The plant is also equipped with a pumping system for water replacement.

Veløykjölpo This 3·5 ha enclosure was completed in 1970, and has a production capacity of between 4 and 600,000 kilos. The enclosed area lies between two islands and has been sealed off by barriers at either end, Fig. 89. Both barriers are formed in concrete with aluminium bar replaceable screens mounted in pairs in guideways. At the south end, Fig. 90, the 30 m barrier was constructed in the dry by bulldozing loose earth and boulders across the entrance, approximately 2 m deep at low water, to form a cofferdam. The aluminium screens are lifted out regularly for cleaning using the light gantry shown on top of the barrier in Fig. 90. At the northern end where the water is 3–4 m deep this method could not be adopted, and concrete pedestals were constructed under water. Unlike Flogøykjölpo, removable aluminium screens in double guideways are used here, so that they may be changed for cleaning, Fig. 91.

When the young salmon reach smolt size at A/S Mowi's fresh water hatcheries, (described in Chapter 6), the smolts are transferred to the salt water enclosures by means of large floating plastic containers. A/S Mowi are very conscious of the fact that the smolts are extremely sensitive at this stage and accordingly have designed a transportation system whereby the fish are never handled.

In the large enclosures the salmon are fed a wet-pellet food consisting of carefully chosen ingredients as close to the salmon's natural diet as possible, with the addition of vitamins. For the feeding, A/S Mowi have designed special feeding automats, which at the same time "pelletise" the wet food.

The managerial control of A/S Mowi's two plants is most impressive. In addition to providing first class feeding, they also ensure that the water quality is as high as possible. To keep the water clean and rich in oxygen, which is a problem at the bottom of deep holes in the enclosures, pumps have been installed to increase the water replacement. Regular checks are made of the dissolved oxygen content at the plants to ensure satisfactory condi-

Fig. 90 Concrete barrier with sliding aluminium screens at south end of Veløykjölpo salmon enclosure in Norway. This barrier was built behind a cofferdam. (*Credit—by courtesy of A/S Mowi*)

Fig. 91 Concrete barrier with sliding aluminium screens at north end of Veløykjölpo salmon enclosure in Norway. The concrete foundations for this barrier were poured under-water. (*Credit—P. H. Milne*)

tions. An examination of the seabed by diving confirmed these rigid controls since there was no sign of either waste food or anaerobic patches on the seabed.

The first 60 tonnes of salmon from A/S Mowi's sea water plants were sold to the market in 1971, not only to home buyers, but also for export to the Continent of Europe.

Bjordal North of Bergen in Sognefjord at Bjordal, Eros Laks A/S has established numerous small, 30 m square, net enclosures round the shores of the fiords. These enclosures are constructed of net on three sides with shore access on the fourth. On the average, the piled net barriers are in 5–6 m of water, with a maximum of 9 m. It is believed that between 25–30 tonnes of salmon were sold to the market in 1971.

Pompano Culture in United States of America

In addition to the facilities described in Chapter 6 for investigating aspects of pompano cultivation,

the Bureau of Commercial Fisheries have constructed several sublittoral pens and enclosures, which are less expensive to build than shore tanks.

The area chosen for a preliminary fish farming experiment in 1967 was a small inlet in Tampa Bay, Florida, Figs. 92, 93 (Finucane, 1968, 1970a). The area enclosed was 2·4 ha and as it was long and narrow only required a short length of netting. This netting was suspended from horizontal timbers supported at regular intervals by timber posts driven into the seabed, Fig. 94. The tides are therefore responsible for the water movement within the enclosure, and the water temperatures vary from a maximum of 32·6°C down to a minimum of 13°C, with a salinity variation of 30·59–36·26‰ (Finucane, 1968).

As well as carrying out growth studies on the fish in the main enclosure, two small sublittoral shore pens, 12·2 m × 12·2 m × 3·0 m were constructed using 6 mm galvanised mesh screens, supported by a timber framework, Fig. 95 (Finucane, 1969). These small pens were used for experiments on the

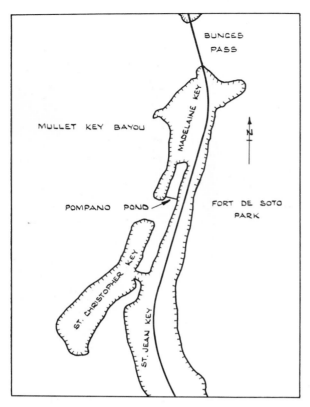

Fig. 92 Location of pompano pond in Fort De Soto Park, in lower Tampa Bay, Florida.

(Credit—From Finucane, 1968)

suitability of different diets for pompano fry and 2,000 fish were stocked in each pen. These diets varied from standard dry fish pellet food to frozen codfish and shrimp meal. Floating food was used so that pond contamination was reduced to a minimum. The floating food is preferred since considerable contamination, with mass mortalities, occurred in previous pompano ponds used for commercial fish farming when trash fish was used for feeding (Finucane, 1970b). The pens were also used to condition juveniles prior to stocking in the main pond.

The advantage of sublittoral enclosures over pond culture is the reduction in capital expenditure at the outset, but the hydrographic conditions of the area must be fully considered, especially the effect of the net enclosure on the environment (Milne, 1971c). Finucane (1971b) reckons that it should be possible to stock pompano at between 123,340–246,900 fish per hectare, depending on water flow, with a possible yearly harvest exceeding 5·6 tonnes per hectare.

Sublittoral Enclosure Research in Scotland

Since 1967 research has been carried out by the Department of Civil Engineering at the University of Strathclyde, into the design and construction of sublittoral net enclosures (Milne, 1969a, 1970b,e, 1972c). This work was sponsored by the British Natural Environment Research Council and enabled full scale structural techniques to be tested in the sea.

Various depths of water call for different construction techniques, and for this purpose enclosure sites have been divided into three categories depending on depth:

(a) shallow water .. .. .. 3–7 m
(b) mid-water .. .. 5–12 m
(c) deep water .. .. 10–20 m

and the types of barriers suitable for each of the above categories are discussed separately.

(a) Net Enclosures for Shallow Water For the purposes of sea farming, an area with a water depth of 3–7 m has been considered as shallow water. For the successful retention of fish and the prevention of damage to the fish net by predators or trash, most enclosures are envisaged with a predator net, hung side by side with the fish retention net. To do this in shallow water, without the snagging and abrasion of the meshes caused by the flood and ebb tides, is difficult without a rigid framework.

An "A"-frame was the first type of framework investigated and stemmed from the British Navy's boom defence experience, where these frames are used in the shallow water portions of boom defence work for the hanging of anti-torpedo nets. There are three general frame sizes using 50 mm galvanised tubular scaffolding, Fig. 96. These frames are held in position by a foot-chain and 910 kg sinkers, as shown in Fig. 97. The tops of the frames are located by a 12·5 cm fine steel wire rope jackstay hung a minimum of 0·6 m above high water, from which the nets are suspended. This type of structure appears very simple as the "A"-frames are assembled on the shore and lowered into place by a barge or other vessel with a derrick, and this presupposes navigable waters. A derrick is also required for the lowering of the sinkers.

This "A"-frame type of framework at first appeared very suitable for adaptation for fish farming purposes, when it was thought that both the predator and retention net could be hung from the fine steel wire rope jackstay with the predator net on the offshore leg and the retention net on the inshore leg. Access to any part of the net could be

Fig. 93   Pompano fish farming site in Fort De Soto Park, in lower Tampa Bay, Florida. (*Credit—U.S. Department of Commerce, National Marine Fisheries Service Biological Laboratory*)

obtained by a wire rope bridge at the apex of the "A"-frame.

However, its construction by naval methods requires assembly on the shore, and installation by a barge with a derrick in navigable waters. As the sheltered coastal sites most suitable for fish enclosures are often not navigable, the underwater construction of the "A"-frame has been investigated, but would appear to be very complex with an inordinate amount of diving. Locations with firm foundations to support the footchains and sinkers where piling is not possible thus would appear to be most suitable for "A"-frame construction.

To overcome the underwater construction problems and placing of "A"-frames, another form of scaffolding manufactured in an "H"-frame was investigated, Fig. 98. This is of standard manufacture in Britain and used to provide 1·5 m wide scaffolding platforms during the cleaning, maintenance and construction of buildings. No clips or

connections are necessary as it is self-locking with one frame every 1·5 m. The bottom frame is the standard "H"-frame, which supports the rest of the scaffolding. Each part is relatively light and can be handled easily, and gives an 0·6 m vertical module.

To test this type of scaffolding, a research project was carried out at the Faery Isles in Loch Sween in 1968, using a collapsible catamaran (Milne, 1970d) in which a construction 9 m long by 1·5 m wide by 5 m high was erected in water of 2 m depth with a tidal range of 1·5 m. To give lateral stability, angled braces were designed using 50 mm diameter galvanised scaffolding tube driven at 3 m centres as shown in Figs. 21, 99. As will be seen from the photographs, the self-lock scaffolding gives a very neat finish. However, a study of the front elevation, Fig. 21, shows a variation in level of the catwalk as the barrier stretches out from the shore. An attempt was made to construct the barrier without divers, by pivoting each "H"-frame into place on the previous one, and then driving a piled leg of 50 mm

Fig. 94   Netting barrier across entrance to pompano pond in Fort De Soto Park, Florida.   (*Credit—U.S. Department of Commerce, National Marine Fisheries Service Biological Laboratory*).

Fig. 95   Shore holding pens and feeding dock at pompano pond in Fort De Soto Park, Florida.   (*Credit—U.S. Department of Commerce, National Marine Fisheries Service Biological Laboratory*)

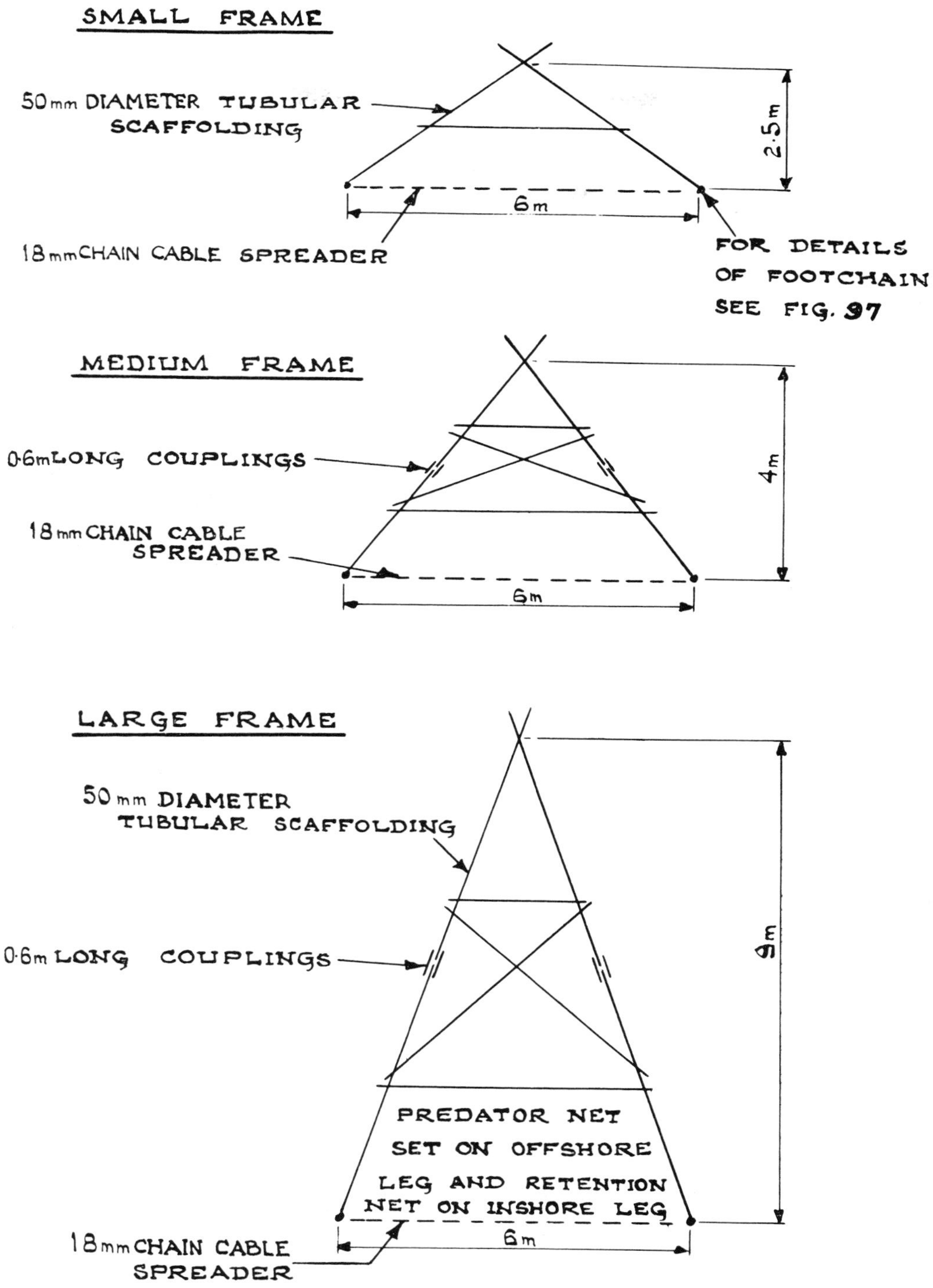

SMALL FRAME

50mm DIAMETER TUBULAR SCAFFOLDING

2.5 m

6 m

18mm CHAIN CABLE SPREADER

FOR DETAILS OF FOOTCHAIN SEE FIG. 97

MEDIUM FRAME

0.6m LONG COUPLINGS

18mm CHAIN CABLE SPREADER

4 m

6 m

LARGE FRAME

50 mm DIAMETER TUBULAR SCAFFOLDING

0.6m LONG COUPLINGS

6 m

PREDATOR NET SET ON OFFSHORE LEG AND RETENTION NET ON INSHORE LEG

18mm CHAIN CABLE SPREADER

6 m

Fig. 96 Fish Net "A"—Frame Barrier—three general sizes. (*Credit—from Milne, 1970e*)

GENERAL LAYOUT OF BARRIER

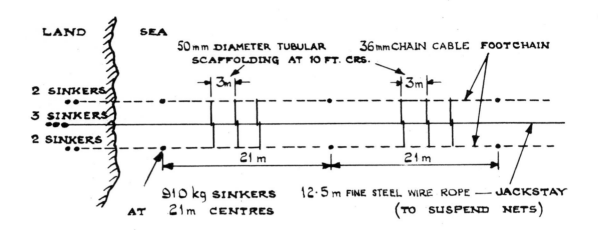

SCAFFOLDING - FOOTCHAIN CONNECTION

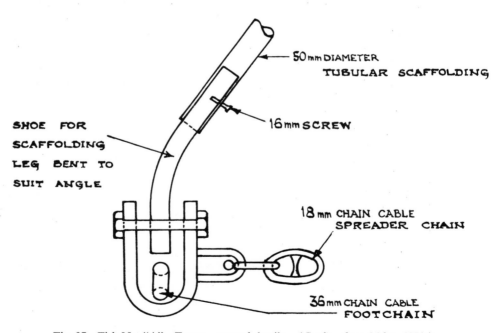

Fig. 97   Fish Net "A"—Frame—general details.   (*Credit—from Milne, 1970e*)

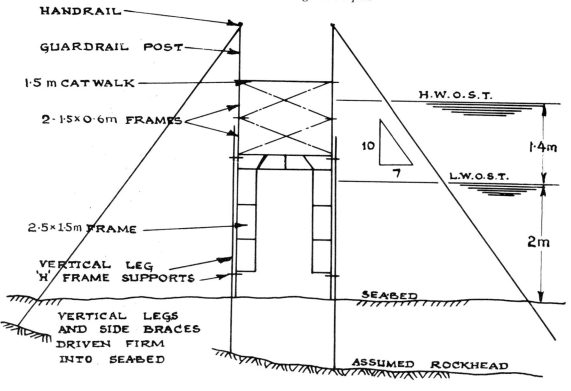

DETAILS FISH NET 'H' FRAME

(MILLS SELF - LOCK SCAFFOLDING SYSTEM)

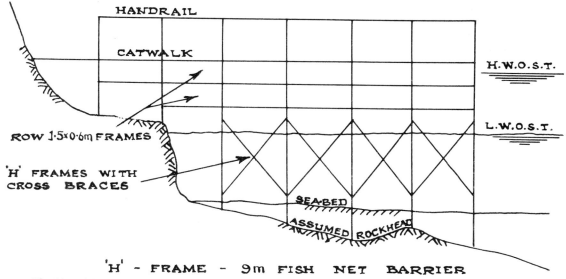

'H' - FRAME - 9m FISH NET BARRIER

Fig. 98 Fish Net "H"—Frame Barrier using interlocking scaffolding tube. (*Credit—from Milne, 1970e*)

Fig. 99 Fish Net "H"—Frame Barrier at Faery Isles, Loch Sween, showing end view. (*Credit—P. H. Milne*)

diameter galvanised scaffolding to give a firm footing in the soft sediment, Fig. 98. This, however, was more difficult than expected and any slight variation from the vertical in one frame is exaggerated as the framework continues by virtue of its interlocking nature. This research project did establish the ease of construction of the self-lock scaffolding, but it requires a construction technique to provide a continuous level footing in either timber or concrete round the periphery, or a technique to drive footing piles with a high degree of alignment.

As a result of the experience gained in the research project on the "H"-frame, it was decided to design a framework relying on driven scaffolding piles which could be connected by horizontal poles set at pre-determined levels for hanging the fish netting. By using a flexible system to be bolted together, due allowance could be made for various depths of penetration of the driven piles (Milne, 1969b).

The hydrographical surveys carried out by the University (Milne, 1972d) have shown that the beds of most sheltered Scottish sea lochs consist of soft sediment so that any structures in shallow water

where nets are used will require piling of some form. To enable small contractors to construct fish enclosures in sheltered waters, where navigable waterways may not exist, a simple design was sought that could be constructed without elaborate equipment or piling vessels.

The "K"-frame design stemmed from the above concepts using 50 mm diameter galvanised scaffolding poles, driven into the seabed using a sledge hammer. To minimise the amount of piling required and to save on scaffolding in deep sediment, the "K"-frame was conceived as shown in Figs. 100, 101. Only two poles require to be driven, the vertical leg supporting the catwalk and fish retention netting and the offshore bracing leg supporting the predator net. As the standard length of poles was 6·4 m the "K"-frame was designed at 3 m centres to simplify construction. Cross bracing was designed for both vertical and offshore legs to give the stability. In the two 12 metres square fish enclosures at the Faery Isles, Figs. 102, 103 (Milne, 1970e, 1972c) the structure was quite stable owing to 2 to 3 m of silt, and the cross bracing for the offshore leg was omitted. Japanese research by Tamura and Yamada (1963) has shown that fish nets should be anchored not only at the top and bottom, but also at the water level. As most Scottish waters are tidal, two poles are required at high and low water. An intermediate pole may be required if the tidal range is greater than 2 m, but in Loch Sween with a tidal range of 1·5 m only high and low water poles were included in the design.

The use of standard 6·4 m long poles, with the design piling module of 3 m centres, simplifies the construction, and the contractors and divers at the Faery Isles were able to erect 6 m of scaffolding per dive, locating all the underwater poles and tightening the connections, Fig. 102. Positive lock couplings for connecting the scaffolding were specified in the design to ensure complete integration of the framework to withstand loadings arising from water, wind and wave action. Screw clip enclosed couplings give a neater finish, but as was found in the construction of Strom Loch Fish Farm described earlier, these clips can slip under load and are not recommended.

As mentioned above, the fish netting should be anchored horizontally at both high and low water as well as at the top and at seabed level. Depending on the type of netting, intermediate horizontal ties will also be required from the spacing given earlier in Appendix 4, that is, at 2 m centres for 25 mm Courlene, 2·5 m centres for 25 mm polythene and 3 m centres for 25 mm galvanised weld-

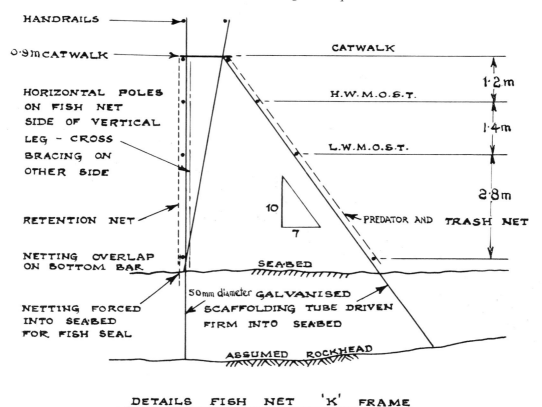

HANDRAILS

0.9 m CATWALK

CATWALK

1.2 m

H.W.M.O.S.T.

1.4 m

HORIZONTAL POLES
ON FISH NET
SIDE OF VERTICAL
LEG - CROSS
BRACING ON
OTHER SIDE

L.W.M.O.S.T.

2.8 m

RETENTION NET

10

7

PREDATOR AND TRASH NET

NETTING OVERLAP
ON BOTTOM BAR

SEABED

50 mm diameter GALVANISED
SCAFFOLDING TUBE DRIVEN
FIRM INTO SEABED

NETTING FORCED
INTO SEABED
FOR FISH SEAL

ASSUMED ROCKHEAD

DETAILS FISH NET 'K' FRAME
(LEVELS GIVEN FOR FAERY ISLES)

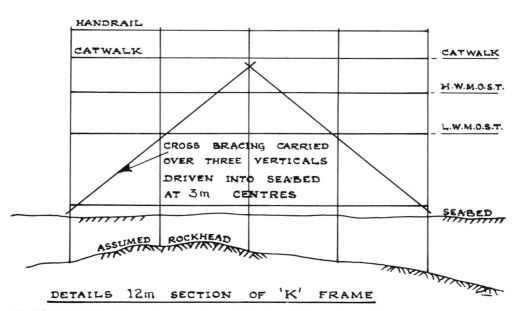

HANDRAIL

CATWALK

CATWALK

H.W.M.O.S.T.

L.W.M.O.S.T.

CROSS BRACING CARRIED
OVER THREE VERTICALS
DRIVEN INTO SEABED
AT 3 m CENTRES

SEABED

ASSUMED ROCKHEAD

DETAILS 12 m SECTION OF 'K' FRAME

Fig. 100   Fish Net "K"—Frame Barrier using individual scaffolding tubes. (*Credit—from Milne, 1970e*)

mesh. With a synthetic fibre mesh net the seabed seal is difficult unless the net is either dug underneath the seabed, or rocks and stones are piled on top to secure it, which makes maintenance and exchange of nets difficult. It is recommended that a rigid mesh is used as a seal at the bottom of the net, to which the other nets can be attached, Fig. 102. These nets can be inserted by divers. Vertical panels of netting are normally the easiest to handle, and those used at Loch Sween were made of galvanised weldmesh in panels of 6 × 2 m.

(b) Net Enclosures for Mid-Water For the purposes of fish farming, an area with a depth of 5–12 m has been considered as mid-water. From 5–7 m there is a choice of a scaffolding framework detailed earlier, or a piled framework. For depths greater than 8 m the scaffolding type of framework is unwieldy and piled structures are the most suitable on which to hang the nets. This piling method has proved popular in Japan (Allen, 1968), and details are given earlier in this chapter of piled structures at Hitsuishi, Figs. 78, 79, Ieshima, Matsumigauru, Megishima, Fig. 82, and Tanoura.

In some cases in deeper water additional restraint to the piles is provided by offshore moorings.

The design of a piled structure to provide a framework from which nets may be suspended to form a fish enclosure requires a knowledge of the underlying sediment. The surface layer should not be taken at face value and cores of the underlying sediment are required to determine their composition and location of bedrock or other stable layer on which to found the structure. In sheltered coastal areas and sea lochs the seabed sediment consists mainly of silt and soft clays near the surface with harder clays in underlying layers down to bedrock. Some sandy seabeds are known but generally in areas exposed to wave action.

To assist prospective marine farmers in appreciating the knowledge necessary for the design of a piled structure some sample calculations are given in Appendix 5, together with some data on the properties of cohesive soils.

(c) Net Enclosures for Deep Water For deep water over 10 m deep, with increasing depth if heavy moorings are required in addition to close pile

Fig. 101 Fish Net "K"—Frame Barrier at Faery Isles, Loch Sween under construction. (*Credit—P. H. Milne*)

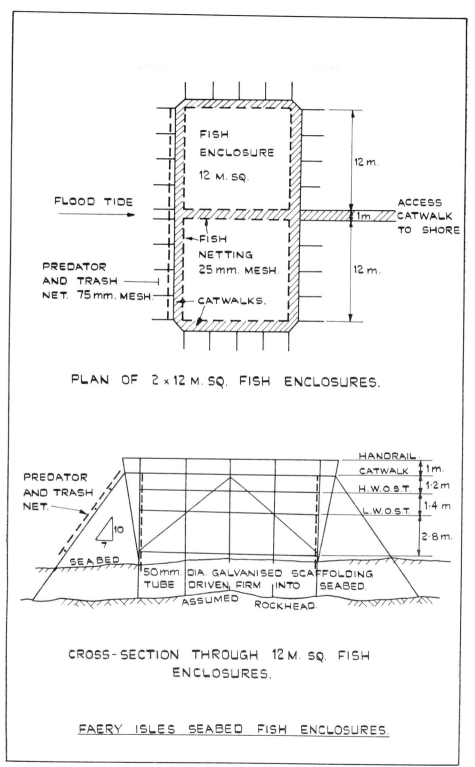

PLAN OF 2 × 12 M. SQ. FISH ENCLOSURES.

CROSS-SECTION THROUGH 12 M. SQ. FISH
ENCLOSURES.

FAERY ISLES SEABED FISH ENCLOSURES.

Fig. 102   Details of two 12 m square fish enclosures at Faery Isles, Loch Sween in Scotland. (*Credit—from Milne, 1972c*)

spacings, it would be more reasonable to install a floating net structure. Also if the seabed sediment is either too hard or too soft for piling it would be necessary to use a floating net. For small enclosures the cost of hiring a piling barge and towing it to a remote site would not be economic and a floating net could be used.

For the purposes of fish farming, an area with a water depth of 10 to 20 m has been considered as deep water. With greater depth it is impracticable to use any form of scaffolding frame or to use piles owing to the close spacing required, and the only practical way to form an enclosure is to hang the nets from surface floats or pontoons.

The only known type of moored nets giving complete closure of the seabed are those used by the British Navy for boom defence. A study was made of their main boom arrangements, but these were rather heavy to be applicable to fish enclosure work since the nets required both large pontoons, and large floats. These consequently required large moorings consisting of sinkers of 8,000 kg weight

and anchors of 9,000 kg. The idea, however, is useful for adaptation for fish net enclosures. As mentioned earlier, both predator and retention nets would be advisable, so a moored support framework has been designed, Fig. 104. Here the nets are hung 2 m apart from the sides of 6 m by 2 m pontoons which are moored at 30 m intervals with one 1,000 kg sinker and one 750 kg anchor to withstand the forces calculated in Appendices 1–4, assuming the use of 25 mm and 75 mm galvanised mesh nets, and a moderate holding ground. The present design (Milne, 1970e) shows a scaffolding and fine steel wire rope framework to keep the retention and predator nets 2 m apart, to prevent snagging and abrasion and to permit cleaning and maintenance of the nets by divers. The use of pontoons enables access to any part of the net, but floats could be used if a tender vessel was always available. On the beach sections between high and low water, pontoons and floats are impracticable and the "A"- and "K"-frame sections already described would have to be used.

Fig. 103   Faery Isles fish enclosures, Loch Sween showing access catwalk.   (*Credit—P. H. Milne*)

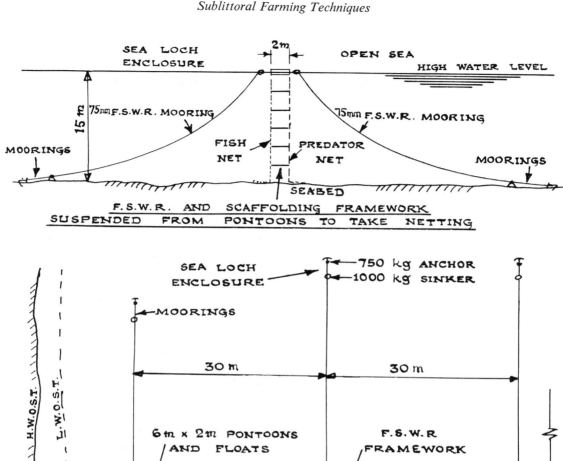

Fig. 104   Moored support framework for netting barrier in deep water.   (*Credit—from Milne, 1970e*)

(d) Boat Access Through Net Enclosures It is inevitable that any net enclosure in coastal waters will interfere with the seaway of small craft. Also it may be necessary for boats to pass through the netting structures, for either stocking or farming the fish, and a boat lock, the same as in use in Japan, would be required. This is formed by two nets 8 m apart supported at the surface by small floats, across a gap of, say, 6 m. All boats using the lock, if fitted with inboard engines, require a protective faring for the propellor; if outboard they must be lifted over the nets. The procedure is to steam up to the first net before shutting off the engine, and allowing the boat to coast over both sets of netting before restarting the engine.

Chapter 9
Seabed Farming Techniques

The seabed has been the traditional zone for the harvesting of shellfish, oysters, mussels, clams and scallops. However, the production of these natural beds can be vastly increased by efficient methods of husbandry and predator control. The transplanting of species to new areas to increase production with the thinning and removal of predators has been carried out the world over. The mechanisation in mussel farming in Holland and oyster farming in the United States are two specific examples of how the production of shellfish can be increased by good management and husbandry. Mention is also made of the first recorded seabed cages designed for lobster farming in Scotland.

(I) MOLLUSCS
Oyster Culture at Long Island, United States of America

Details are given in Chapter 6 of the hatchery techniques for the cultivation of the American Oyster, *Crassostrea virginica*. Once the oyster spat are ready for planting on the oyster beds the first year of bottom culture is the most critical. One of the chief problems during this period is silting, since any covering of silt interferes with the feeding, thereby arresting the growth, and at worst may smother the oyster. The selection of sites with a hard bottom and strong tidal currents helps to prevent this. It has been found that the beds must still be kept clean of silt by regular suction dredging. To prevent a build-up of silt new techniques are being developed to clean the silt from the oysters at regular intervals with water jets operated from work boats. However, despite these precautions it is still wise to move the young oysters for harvesting and transplanting, at least once and preferably twice during their first year. Visual examination of the oyster beds is the best method to prevent problems arising and four of the personnel of G. Vanderborgh and Son are qualified SCUBA divers and are employed for this purpose to regularly monitor

the oyster beds approximately twice a week (Webber, 1968).

Initially the seed oysters from the Vanderborgh hatchery are planted on the beds at a density of 1,250 bushels (1 bu = 35·2 l) per hectare. At the end of the first year the oysters have grown to 12,500 bushels and must be distributed over 5 ha. By the end of the third year they have grown to 25,000 bushels and must again be replanted over 10 ha. The annual production from this planting density and growth rate is equivalent to some 1,250 bushels per hectare. These production estimates are well in excess of the present average production figures for other oyster grounds in the United States. For comparison they are 30 times the yields from other leased oyster beds and 600 times the yields of public oyster beds.

The extent of the areas required for oyster culture can be gauged from that fact that G. Vanderborgh and Son and the Radel Company currently lease 1,040 ha of oyster beds from the towns of Oyster Bay and Northport on Long Island Sound. These beds are in 1–12 m of water and the techniques involved in planting and the subsequent care of the oysters in these beds represents a major advance in oyster culture in North East America (Ryther and Bardach, 1968). Long Island Oyster Farms, Inc., one of the other commercial oyster growers in Long Island Sound also leases 600 ha of oyster beds.

Oyster Culture in South Africa

Oyster cultivation was South Africa's first marine farming venture, and was started several years ago by the Knysna Oyster Co. (Anon., 1971d). Unfortunately the project has not yet overcome all the problems of commercial production, because despite using specially prepared abalone shells for the spat-fall, the settlement has been very poor during the last two years. It is thought that the reason for this has been very low sea temperatures off the Knysna coast during the spawning season which has depressed the spat-fall.

Normally after collecting the spat on the shells the young oysters are transplanted to Belvidere Bay in the Knysna Lagoon. After two years these cultivated oysters are gathered for harvesting, and those too small for marketing are left out in trays for a further year. In 1970 the Kynsna Oyster Co. sold over 40,000 Belvidere oysters. It is hoped in the future to build an oyster hatchery to provide a regular supply of spat for transplanting.

Mussel Culture in Holland

Holland is the world's second largest producer of mussels with an annual total of 100,000 tonnes (Mason, 1971). The main area for mussel culture, *Mytilus edulis*, is the Waddensea where mussels are transferred from areas of great abundance and overcrowding to areas of good growth and fattening potential. Here the farmers are allocated areas of the seabed for cultivation. Using large dredges, the spat are collected from public grounds and are then spread over the farm or "park", which may vary from 5 to 18 ha in area. Before transplanting the farmers clear the leased areas of unwanted predators, especially starfish, using a special "roller" dredge. When spreading the spat, the farmers generally break up any large clusters to ensure maximum growth (Havinga, 1956, 1964). The mussels are subsequently thinned to ensure better growing conditions before marketing.

The bottom cultivation of mussels in the Waddensea with its small mean tidal range of 1·5 m has allowed a high degree of mechanisation, which has increased the yield and consequently lowered the price. With the majority of the "parks" in the sublittoral zone, over 200 boats, each 15–20 m long, are employed in mussel farming, each with a capacity of 40 tonnes or more. The chief advantage of this bottom technique is that the mussels are submerged for the whole of the tidal cycle and thus their feeding period is not interrupted as in the intertidal zone.

Mussel Culture in Ireland

The edible European mussel, *Mytilus edulis*, spawns naturally along the Irish coast, and mussels have been harvested for many years from the natural beds in Castlemaine Harbour, the River Boyne Estuary, Wexford Harbour, Dundalk Bay and Carlingford Lough (Meaney, 1970). However, as many of these mussels are small and of poor condition recent investigations have been carried out by

the Irish Sea Fisheries Board to assist in the cultivation of mussels in areas with the necessary conditions for good growth and development.

To improve the mussel harvest the transplanting of mussels to growing areas is now practised in Castlemaine Harbour where mussels settle extensively. In 1968, over 4,000 tonnes of mussels were relaid in areas for intensive cultivation where they are regularly tended by fishermen to prevent predation by starfish and crabs. More recently the transplanting of mussels has been extended to other areas and in 1969 over 150 tonnes were relaid in Carlingford Lough and 200 tonnes in Wexford Harbour (Edwards, 1970). In 1970 there was a commercial transplant at Carlingford Lough of 500 tonnes and a trial transplant of 100 tonnes at Cork.

In addition to these investigations comprehensive surveys of all the suitable bays and estuaries round the Irish coast are under way to assess potential development areas. Also included are diving surveys to study the nature of the seabed and the seed potential. The possibility of raft culture is also being considered, as mentioned in the next chapter.

Clam Culture in United States of America

Both the hard clam, *Mercenaria mercenaria*, and the soft clam, *Mya arenaria*, are suitable for cultivation.

In 1962 about 22% of the hard clams sold on the east coast of the United States were farmed on private grounds (Iversen, 1968).

Most of the procedures and techniques described for oyster culture are also applicable for clam cultivation. The main requirement in the growing area is a firm bottom in which the clam can burrow, and a regular flow of water to provide the clam with plankton for its development. Unlike oyster farming, collectors are not required since the clams will settle directly on the bottom. Predators such as starfish and crabs are, however, a problem, and planted beds have to be regularly inspected to prevent losses. Harvesting requires either a hydraulic dredger to loosen the bottom to suck up the clams (Kerr, 1970), or the old traditional hoe in the intertidal zone.

To try and increase production many attempts have been made to protect the clams from predators using such techniques as screened boxes and trays, fenced enclosures, covered netting, and salt water ponds and tanks. However, these methods are expensive and require extensive maintenance since they can cause silting and slow growth.

A new method developed at the Virginia Institute of Marine Science (VIMS) involves spreading shell, gravel or aggregate over the seabed prior to planting the seed clams (Castagna, 1970). This aggregate should be spread to a depth of 25 to 75 mm over the seabed, and can be placed at any time of year. However, the seed clams should only be placed when they are active, normally when the water temperature is above 9°C, since the clams will burrow under the aggregate in a shorter time at higher temperatures. This layer of aggregate protects the clams, and VIMS report that over 80 % of the seed clams will survive even when predators such as starfish and crabs are present.

The development of this cheap method of protection will not only boost commercial production, but also encourage the operation of clam hatcheries, which can now use artificial foods for feeding the larvae (Chanley and Normandin, 1967).

Scallop Culture in Russia
An investigation into artificial cultivation methods for edible scallops has been conducted by the laboratory of Marine Investigations of the Zoological Institute of the Academy of Sciences of the U.S.S.R. The research was carried out on *Mizuhopecten yessoensis* and *Spisula sachalinensis* at Posjet Bay on the Western shores of the Sea of Japan (Golikov and Scarlato, 1970).

A study of the bottom distribution of *M. yessoensis* using skin divers showed them to inhabit water depths of 0·5 to 25 m on sand or silty sand bottoms. In summer the scallops come inshore and are most abundant in depths of 1·5 to 4 m at temperatures of 15–20°C. However, in autumn before the frosts they move offshore down to 25 m, being most abundant in the 6–8 m zone. Spawning of *M. yessoensis* starts in May and continues until August with the young spat settling on fronds of seaweed in depths of 1–3 m with 7–15 specimens per m². However, on artificial collectors such as cotton nets, sisal ropes or birch brooms suspended from rafts in depths of 1–3 m, 1,000 to 2,000 larvae settle per m². After about a month the young molluscs detach themselves to begin their bottom mode of life. However, many are lost to predators such as starfish. The artificial cultivation of *M. yessoensis* may thus be accomplished by firstly providing a suitable settling collector of some fibrous material suspended from a raft at depths of 1–3 m, and secondly providing a suitable substratum protected from predators, such as in Netlon mesh bags (Nortene, 1969) suspended from rafts or moored on racks off the bottom.

Studies of the scallop, *S. sachalinensis*, showed it to be most prolific in depths of 1–5 m on sandy bottoms. The mature specimens spawn in June, the spat settling on the bottom, and by the end of July the young molluscs have a shell height of 2–3 mm. The limiting factor in nature to the abundance of *S. sachalinensis* is the mortalities in the autumn during storms, when heavy waves often cast them ashore, before they are old enough to bury themselves into the seabed. The artificial cultivation of *S. sachalinensis* could therefore be carried out by providing suitable bottom collectors for the spat and then transferring the collectors to protected farm areas which favour further growth.

(II) CRUSTACEANS

Lobster Culture in Scotland
The European lobster, *Homarus vulgaris*, normally requires about six years in the sea before it reaches the minimum marketable length of 23 cm, which is a long time for a commercial farmer. Dr H. J. Thomas (1964) of the Department of Agriculture and Fisheries for Scotland has worked with lobsters for many years and breaks down the commercial rearing possibilities into three avenues:

(i) The short-term rearing of lobsters by hatcheries from the egg to early stages for release into the sea;

(ii) The long-term rearing of lobsters by raising the young up to marketable size in captivity from a hatchery spawning stock; and

(iii) The short-term retention of lobsters for growing and fattening.

Of these three methods, (i) has never been shown to be financially successful since the cost of rearing the young is well in excess of the value of the lobsters recaptured. Method (ii) is only profitable if the rate of growth can be increased by using warm water to reduce the number of years to achieve market size, and research has been carried out along these lines in the United States (Kensler, 1970). The third method, (iii) at present is the only one which is commercially carried out economically (McKee, 1967), and one such enterprise has been set up in Scotland.

In 1968 Lieut.-Commander J. Futcher established a lobster farming company at Kinlochbervie in Sutherland near Cape Wrath. The immediate operation of the Kinlochbervie Shellfish Company is to purchase and capture marketable size lobsters in the summer, Fig. 105, for growing and fattening for several months before sale during the winter when prices reach their peak (Milne, 1970c). Futcher chose seabed cages for retention of the lobsters.

The design of seabed cages is much simpler in concept than either surface or submersible cages since no flotation or buoyancy compartments are necessary, although divers are required for construction. Also, since the structure rests on the seabed, outwith the range of wave and wind action, a seabed cage can be constructed from much simpler and lighter materials (Milne, 1972c). The first seabed cages for lobster holding were 1·8 m × 1·2 m × 0·5 m, made from galvanised weldmesh, placed on sandy bottoms in Lochs, Laxford and Inchard (Mundey, 1969). Futcher found, however, that the stocking, feeding and harvesting of such small units was awkward since this could only be carried out by divers. To simplify these problems a large cage

with a capacity of 2,250 kg of lobsters was designed in 1970. These cages were made from synthetic fishing net, 12 m × 12 m × 2·4 m and were anchored on the seabed in 1970 in depths between 10–15 m.

The author was the company's first visitor to these seabed cages in 1970 (Milne, 1971c) when he dived to study the hydrography, water and sediment movement in the vicinity of the cages. By carefully selecting seabed sites with good water movement, yet not exposed to storm conditions, no problems should arise due to insufficient water movement. In 1969 the Kinlochbervie Shellfish Co. sold 4 tonnes of lobsters and this was raised to 12 tonnes in 1970 and 1971. It is hoped to achieve an annual output of 20 tonnes using this method (Bowbeer, 1971).

Fig. 105 Lobsters, after capture at Kinlochbervie in Sutherland, are held overnight in a floating storage box, prior to shipment by rubber dinghy the next day to the seabed lobster cages in Loch Inchard. Lt.-Cmdr. J. Futcher, Managing Director of Kinlochbervie Shellfish Co. on far right. (*Credit—P. H. Milne*)

Chapter 10

Floating Cages and Rafts

The development of new floating techniques for the farming of marine organisms has increased the potential development of many areas throughout the world. The two main advantages of floating techniques are firstly that they can be used in areas where the seabed is unsuitable for traditional shellfish farming, and secondly by using an off-bottom method the predators can be more easily controlled with less loss of stock.

The two biggest developments in this field have been made in Japan with the floating cage cultivation of yellow-tail, and in Spain with the floating raft cultivation of mussels. These floating techniques with minimal capital construction costs now enable small crofting communities, as mentioned in Japan, to participate in the development of sea farming.

The design and mooring considerations for the installation of surface facilities is mentioned with the difficulty of access to individual cages. The grouping of cages round a spider framework, or on a large raft, simplifies both the feeding and maintenance of the unit.

(I) MOLLUSCS

Oyster Culture in Japan

The cultivation of oyster seed for farming in Japan was discussed earlier in Chapter 7. Once there is a good settlement of oyster spat, *Crassostrea gigas*, on the collecting rens the oysters are ready for suspension in growing areas for marketing. New rens are made up depending on the water depth at the site, and in the Inland Sea some attain 10–15 metres in length. Two floating techniques are used—either the raft or long-line.

The floating rafts used for oyster culture in the Inland Sea are of a fairly standard size, 16 m × 8 m, carrying between 500–600 wire rens, Fig. 106. The rafts are of simple construction, using 10–15 cm diameter bamboo (sometimes cedar) poles which are lashed together with wire in two layers at right angles to one another with the poles 0·3–0·6 m apart (Fujiya, 1970). These frameworks are supported by hollow concrete drums, tarred wooden floats, or styrofoam cylinders. All new rafts being constructed are using the new styrofoam cylinders, which in some cases where marine fouling is bad, are encased in a polythene bag for protection. As the oyster rens increase in weight more floats are added to the raft to maintain stability. These rafts are generally moored in rows, 5–10 m apart and tied together. The number of rafts on one line varies from two to ten with anchors at both ends. The rafts are generally left in the water throughout the year, but may be brought inshore in the spring for maintenance, since they are relatively crudely constructed using local materials and only have a life of five years.

The long-line method is a modification of the above hanging method and was first applied in Japan in 1947. The standard long-line is 70–75 m long, and consists of a parallel pair of 2·5 cm diameter lines buoyed up by tarred wooden barrels or styrofoam floats. The lines are run through eyelets at the end of each float which at the beginning are spaced at 3 m apart. Later when the wire rens are weighted with oysters this spacing is reduced to 1·5 m. The ends of each of these lines of floats are secured by two or three anchors, depending on the location. Sometimes a series of long-lines are strung in parallel 10 m apart and the centre of the line is also anchored to prevent snagging of adjacent lines. The vertical rens are then suspended at 30 cm intervals along the lines. The popularity of the long-line method is due partly to the lower cost and maintenance than the raft technique. More important is the fact that the long-line method can withstand winds, waves and currents better than rafts, and can thus be used on more open coastlines. This method has therefore opened up new areas for oyster culture round the Japanese coastline not suitable for rafts.

The growth rate of the oysters depends on the environmental conditions at the site such as water temperature and available food. Thus oyster farming falls into two categories, one- and two-year farming. In the one-year farming method, the oysters are transferred to rafts immediately after settlement in the spring and are harvested the following spring. In the two-year method the oysters are kept in shallow water for the first year and are transferred to rafts for further growth the second year before harvesting the following spring, two years from settlement. The advantages of two-year farming are apparent when one considers production figures per oyster of 5–10 g of shucked meat in one year and 10–30 g of shucked meat in two-year farming. However, one requires suitable conditions to keep the oysters for two years without mortalities and hence the hardening process in shallow water during the first year.

The total annual production of oyster meat from Japan since 1963 has remained around 35,000 tonnes, most of which is consumed locally. This total is nearly double the production total of 19,000 tonnes in 1957 and shows the expansion and development of oyster culture in Japan (Ryther and Bardach, 1968). In this the Inland Sea is extremely well sheltered and protected from the weather, permitting raft culture with the ability to use at least 10 m of the water column for culture, which is very advantageous.

Oyster Culture in Spain

Following the success of mussel raft culture techniques in Spain, the attention of Spanish industrial firms and scientific institutions has now turned to the cultivation of the European flat oyster, *Ostrea edulis*. Since the natural beds of oysters have long been exhausted in Spain, it has been necessary to import oyster spat from Brittany. At present oyster cultivation is being carried out on floating rafts in Galicia (Andreu, 1968a,b; Figueras, 1970) using

Fig. 106 Floating raft for oyster culture in Japan, using bamboo poles and styrofoam cylinders. (*Credit—by courtesy of the White Fish Authority*)

Fig. 107 Fibreglass buoyancy collar for oyster culture in Nova Scotia. These collars measure 2·4 m × 1·2 m.
(Credit—by courtesy of World Fishing)

the same raft techniques discussed elsewhere in this chapter for Spanish mussel culture.

Attempts had been made originally to cultivate oysters on the seabed but the bottom turned out to be too muddy. The raft culture of oysters is now recommended for growing and fattening, with ropes suspended vertically from the raft framework. Some others are maintained in suspended mesh bags of Netlon (Nortene, 1969) but this is not common as the technique is only in its infancy. During cultivation the oysters are handled three times. Initially the oysters are attached to the rope with cement before hanging from the raft. Then they are cleaned during August and early September to remove seaweed and other fouling organisms which would impede growth. Lastly they are detached from the ropes in November for harvesting and cleaned once again.

Due to lack of natural oyster spat attempts are being made to recover areas of the muddy seabed for oyster culture by dumping old shells and gravel to stabilise the bottom. Mature oysters placed in Netlon bags have then been placed on these beds, and using conventional spat collectors of limed tiles (a method used in Brittany), satisfactory results have so far been obtained, which indicate the promising nature of this technique. Meantime oyster spat will need to be imported, and at present 10 million oysters are maintained in Galicia.

Oyster Culture in Canada

Since 1966 trials have been under way in Nova Scotia

into off-bottom techniques for oyster culture due to the unsuitability of the seabed in many areas. In fact, using the traditional methods of transplanting the oysters on the bottom in the Maritimes, due to predation by starfish, silting and smothering, often only 2% of those planted survived.

The area selected for these trials was Pleasant Point, Halifax County, where there had once been a natural population of oysters. The Department of Fisheries who are conducting the trials are currently testing various methods of flotation, including:

    (i) flotation collars of styrofoam covered with aluminium,
    (ii) plywood reinforced with fibreglass,
    (iii) fibreglass filled with polyurethane, and
    (iv) fibreglass.

These collars measure 2·4 m × 1·2 m, Fig. 107, and have sufficient buoyancy to float five oyster trays, 2·3 m × 1·1 m, giving a total artificial bottom of 12·5 m² (Anon., 1971c).

Some three-year old oysters from Pleasant Point were recently subjected to a market appraisal, and reassuringly were found to be superior in flesh and flavour to the bottom oysters from Maritimes.

With the establishment of a new oyster hatchery at Pleasant Point, some other salt water areas are now under investigation into their suitability for raising oysters using this off-bottom method.

Oyster Culture in Norway

Norwegian waters offer excellent natural conditions for oyster breeding and yield a high-grade oyster spat, *Ostrea edulis*, which can survive well even with large temperature fluctuations. This is most important in Norway due to the high summer and low winter temperatures in the Norwegian fiords.

Low salinities in the surface water, and little oxygen in the bottom waters due to lack of circulation means that oysters can only be cultivated at mid-depth (Iversen, 1968). Here cages are suspended from the surface to maintain the oysters in suitable conditions of oxygen, food, temperature and salinity for growth. Even then it still takes three years to produce marketable size oysters, which is an expensive business since considerable labour is required to maintain the cages. However, the excellent condition of the high-grade spat with exports to Britain, Denmark and France has kept this industry alive. The biggest oyster farm in Norway is owned by A/S Østers at Vagsstranda near Aalesund, and in addition to raising 200,000 oysters a year for home consumption, also exports spat. In 1971 some 3 tonnes of baby oysters were exported to Britain (Anon., 1971m).

Oyster Culture in Venezuela

An oyster culture programme has been underway since 1968 at Isla Margarita, Punta de Piedras. This work has been carried out by the Marine Research Station of the La Salle Foundation (Anon., 1969a). The results obtained indicated that the spawning and spatfall of the mangrove oyster, *Crassostrea rhizophorae*, are continuous, with maximum spatfall from June to September.

Three types of collectors were tested, and the best results were achieved on wooden planks painted with bitumen, similar to the Australian technique of tarred sticks. Collectors were placed both in the mangrove zone and also on a floating raft. The predators normally present in the mangrove area could not affect the floating raft, and hence higher spatfalls were achieved on the raft. Unfortunately marine fouling by bryozoans, tunicates and algae is quite serious in this area, but these organisms were eliminated by lifting and exposing the collectors occasionally to sunlight for 30–60 minutes.

The observed average rate of growth of the oysters has been 10 mm per month, indicating two harvests per year or possibly three during a two-year period.

Mussel Culture in Spain

The cultivation of mussels, *Mytilus edulis*, started in Spain in the early nineteen hundreds using the French "bouchot" system described in Chapter 7.

However, this system did not produce particularly good mussels and it was not until the Japanese raft technique for oyster culture was applied to mussels and tested on the Galician coast of northwest Spain in 1946 that any further progress was made. Since then the industry has progressed to become one of Spain's most important industries with an annual production total in 1968 of 140,000 tonnes (Andreu, 1968a,b).

Although the floating technique of mussel culture was first tried on the Mediterranean coast at Tarragona and Barcelona early this century, it was not until the possibilities of the north-west coast of Spain were considered did production attain any importance. Original experimental raft cultivation was first carried out by Dr B. Andreu, Director of the Laboratorio del Instituto de Investigaciones Pesgueras, of Vigo in Galicia, and it is to him that much of the credit is due for Spain now becoming the world's leading producer of mussels (Andreu, 1968a,b; Mason, 1971).

The Galician rias selected for mussel culture are similar to the Norwegian fiords, in that they are long, up to 25 km, and narrow, 3–12 km wide, with depths ranging from 30 to 60 m. Raft culture is, however, confined to areas 3–10 m deep due to the problems of mooring. The bays in general have narrow entrances, and are well protected from winds and storms. Although in a sheltered area, the salinity is oceanic (approximately 35‰) with summer temperatures of 20°C only falling to 10°C in winter. Also due to the 4 m tidal range, the tidal currents generated bring in food for the mussels. Because of the hanging raft culture technique used there is little danger from predators, and care is necessary to ensure the ends of the ropes do not touch the seabed, since the mussels could be attacked by starfish.

Initially the rafts were constructed from old ships, and a wooden framework was suspended over the side above the water level, held fast by backstays to the mast, Fig. 15. To prevent marine fouling, the ships' hulls were treated with cement. Subsequently instead of ships' hulls, a central float was used in the form of a large box, with a trap door for maintenance and repairs. The beams for the collecting ropes were then shored up with backstays from 2–4 masts in the middle of the float (Wiborg and Bøhle, 1968). Modern improvements to these original rafts have come in the use of 2–6 smaller floats to spread out the weight, and then backstays are not required. Early floats were made with a timber and wire mesh framework covered by concrete, but recently fibreglass floats are beginning to replace the original concrete floats (Mason, 1971). The average raft is 20 m × 20 m and will accommodate up to 500 suspended ropes. Larger rafts now being built with more streamlining for exposed locations can handle up to 1,000 ropes (Ryther and Bardach, 1968).

The original ropes used to collect the spat were made of loosely-woven esparto grass (made locally), to a diameter of 3 cm. More recently these are being replaced by a thinner 1·5 cm diameter nylon rope. The length of these ropes varies from 3 to 12 m depending on the depth of water below the raft at low tide. Before use, small wooden pegs 1·5 cm square and 25–30 cm long are inserted between the strands of the rope at intervals of 40 cm to prevent clumps of mussels sliding down. After the pegs are inserted the whole rope is tarred, and Mason (1971) reports that if this is done periodically the ropes can last for up to 10 years.

The rope collectors are put out in the Galician bays in April to collect the May settlement and by the end of the autumn have grown to 30–40 mm. By the second autumn they have reached 80–100 mm and are ready for marketing. Due to the rapid growth on the ropes, some thinning is necessary, and the outer layers of mussels are periodically stripped off and attached to new ropes by binding them round with cotton netting which eventually rots, by which time the mussels have reattached themselves to the rope by their byssus thread. This transplanting or thinning of mussels is best done on dull days since bright sunshine hinders the byssus attachment. Although mussels are said to favour water with a high light intensity, Paz-Andrade (1968) reports that the intense summer sunlight however inhibits growth, and a 69% increase in weight can be obtained by growing the mussels in shaded conditions. Thus production can be increased by placing sunscreens over the timber frameworks from which the ropes are suspended. Under normal conditions a 10 m long rope can be expected to produce on the average 120 kg of mussels annually. Therefore a 500-rope raft can produce 60 tonnes per year, and the annual yield per hectare (approximately 10 rafts) may be estimated at 600 tonnes. With a total of over 3,000 rafts now in Spain (94% in Galicia), the total potential is 150,000 tonnes. This, however, is not the production limit, since it is only the necessity of finding new markets that is limiting expansion.

Mussel Culture in Venezuela

The cultivation of the local Venezuelan mussel, *Perna perna*, has been attempted using the same techniques developed in Spain, described previously. At the outset Spanish scientists went to Venezuela to assist in building the rafts and start mussel farming (Iversen, 1968). The first rafts, 7 m square, were constructed from bamboo poles and styrofoam floats. Bamboo poles were used as spat collectors, as discussed earlier for the Philippines. Initial results were very promising since the mussels grew very rapidly with favourable meat to total weight ratio. Unfortunately, however, the first large scale farming venture of 122 rafts was unsuccessful since the bamboo poles became riddled with marine borers, and broke up.

The presence of marine borers, and other fouling organisms should obviously be considered at the outset before material selection. New studies are now being launched, and smaller rafts are being constructed to allow fishermen with little capital to participate in local mussel farming in the hope that it will become profitable.

Mussel Culture in Norway

The success of the Spanish floating raft culture, discussed previously, prompted research into the possibility of cultivating mussels using the Spanish system in Norwegian fiords. This research was initiated by the Havforskningsinstitutt (Marine Research Institute) in Oslofiord (Bøhle and Wiborg, 1967). However, the early experiments were not successful since the weight on the spat collectors was too great, and the shell clusters gradually became too heavy and fell off.

In Norway some of the inlets have high summer temperatures and in these the mussel spat, *Mytilus edulis*, which settle in the spring grow to approximately 30 mm by the autumn. This size is, however, considered too small for thinning employing the Spanish technique of using fine cotton netting, when a size of 40–50 mm is desirable. To overcome this problem a net bag has been designed from polypropylene fibre with a mesh width of 6·5 mm × 12 mm and a diameter of 3·0 cm. This net bag can then be filled with the small spat via a funnel for subsequent raft suspension (Bøhle, 1970). Using these bags a harvestable mussel 50–75 mm in length can be obtained by the second autumn.

Bøhle (1970), reports that the preliminary results at Melsomvik and Nordåsvatnet have shown the benefits of this system. In future the use of net bags for transplanting spat will enable spat to be moved to—(a) areas where there were no mussels before, but where growing conditions are good, and, (b) areas where winter ice would interfere with floating structures, but summer conditions were suitable.

Mussel Culture in Scotland

Although some 400–500 tonnes of mussels are collected annually from the Scottish shoreline, few are for human consumption, and they are mainly used for bait in line fishing. However, since mussels, *Mytilus edulis*, lend themselves to economic cultivation for the table market, and some of the west coast sea lochs are similar to the Spanish rias and Norwegian fiords with high summer temperatures, a research programme was initiated at the Marine Laboratory, Aberdeen, of the Department of Agriculture and Fisheries for Scotland (Mason, 1967), using raft culture techniques.

Initial research was centred at Loch Tournaig, off Loch Ewe in Wester Ross, adjacent to an existing marine field station. However, the water temperatures in this area did not approach that of the Spanish rias where summer temperatures reach 20°C and only fall to 10°C in winter. The rate of growth of the mussels was good considering the conditions, and Mason (1969) suggests that commercial size mussels could be grown over a period of three years.

The only area off the west coast of Scotland with comparable summer temperatures to the Spanish rias is the Sound of Jura, due to the extremely small tidal range 1·5 m at springs, Fig. 18. Temperature observations in the sea lochs in this area,

especially in Loch Sween, have regularly reached 20°C due to their sheltered nature and small tidal range, which gives higher temperatures than the nearby coastal waters (Milne, 1972d). Linne Mhuirich, a small inlet (4·4 km long by 0·4 km wide) off Loch Sween was the second site chosen by Mason (1969) for raft mussel trials, Fig. 108, due to its restricted narrow entrance which reduces the tidal range to 0·9 m at spring tides.

The spawning of mussels in the Loch Sween area occurs in the spring with spat fall starting in May or June, reaching the maximum in July and August (Milne and Powell, 1972). At the outset three types of rope, coir, sisal and Courlene were tested for comparison as to their suitability for settlement. The most fibrous, the coir, gave the best settlement. To prevent the clumps of mussels sliding down the ropes, small pegs were inserted into the rope, similar to those used in Spain, since the clumps soon reached 30 cm in diameter, Fig. 4. The results of the trial experiments showed that the size of the mussels from spring to autumn reached a mean of 43 mm. By the following July, one year after settlement, they measured up to 70 mm, and by September were up to 75 mm, a marketable size. Mason (1969), says that these growth rates are far greater than any recorded for naturally occurring mussels in Britain,

Fig. 108   Experimental raft used for the rope culture of mussels in Linne Mhuirich in Loch Sween, Scotland.

(*Credit—P. H. Milne*)

and compare favourably with those recorded anywhere in Europe, with the exception of the Galician raft mussel culture. A 3 m rope can thus produce at least 25 kg of mussels in 18 months from settlement, that is from the spring to the following autumn.

The success of these raft trials has prompted a pilot commercial unit to be set up in Linne Mhuirich by Mr Tom Stevenson. Due to the larger moorings required for rafts, he is using individually buoyed ropes in groups, Fig. 109, and buoyed frames, Fig. 110, where up to eighteen 3 m ropes can be suspended from each frame (Milne, 1970c). The initial reaction to these commercial mussels has been good and it is intended to increase production to 30 tonnes per annum.

Mussel Culture in Ireland

In addition to the recent transplanting of the edible European mussel, *Mytilus edulis*, described in the last chapter, investigations are also under way into the feasibility of raft culture in Ireland.

Encouraged by the results of research into mussel culture using the Spanish raft system in Scottish west coast sea lochs (Mason, 1969), it was thought probable that many of the inlets on the west coast of Ireland might also be suitable. This work is being carried out by University College at Galway (Murray, 1971).

Using rafts constructed of timber and expanded polystyrene for flotation, the settlement, growth, condition and yield of mussels grown on ropes has been investigated. The ropes were immersed prior to the June 1969 settlement, and the mussels grew to a mean length of 57 mm in 10 months with an average production of 6·9 kg/m of rope.

Mussel Culture in Germany

Following the success of the raft culture technique in the Spanish rias, recent experiments have been carried out adopting similar methods at Flensburger Förde in the Baltic Sea (Meixner, 1971). Long narrow rafts, 3·8 m × 0·8 m were built from polyethylene pipes 25 cm in diameter, and ropes of different materials were used as the spat collectors for the larvae of *Mytilus edulis*. The first ropes were suspended from the raft in the spring of 1969 and the newly metamorphosed larvae settled over the period, June to September. By December the mussels varied in size from 1·5–3·5 cm, and the ropes were removed from the raft and taken ashore during the winter period of ice cover. The mussels were put out again on the raft in April 1970, using long net bags of polyethylene.

Fig. 109 String of buoys, each supporting a 3 m long rope for mussel culture at Linne Mhuirich in Loch Sween, Scotland. (*Credit—P. H. Milne*)

These bags, 2·5 m long, 20 cm diameter, with a mesh size of 1–2 cm, were filled with mussels and tied at intervals of 30–40 cm. By the end of the second summer the mussels varied in length from 4·5–6·5 cm, and when weighed, produced 30–50% more meat than mussels of comparable size from the seabed. These tests have shown that the raft culture technique for mussels in the Baltic can produce a better product than similar seabed culture, but it remains to be seen if the economics of the extra handling and over-wintering on land are practicable.

Mussel Culture in Australia

A small quantity of the edible mussel, *Mytilus edulis planulatus*, is harvested each year as a side-line of the Australian scallop industry, but these are usually of poor quality. Investigations into the feasibility of commercial mussel growing are now being conducted in Sydney Harbour by Dr R. J. MacIntyre of the University of New South Wales in Sydney (Anon., 1971i), since cultured mussels are expected to grow faster and be of superior condition.

Since the mussels grow naturally in the bays and estuaries it is simply a matter of suspending a suitable surface for spat settlement. The method adopted has been a development of the Spanish raft mussel culture, described earlier in this chapter. The suspension of ropes is favoured in Australia to avoid attack by whelks, borers and starfish which can cause considerable damage.

The experimental raft construction used is very simple, consisting of a module of three timber battens, 7·5 cm × 5 cm, supported by two 200 litre drums, Fig. 111. Up to six of these modules can be moored together using old motor tyres as spacers, the whole raft being linked by a peripheral chain joined to a single anchor point at the bow, Fig. 112. Hardwood timber is used for the battens and these are joined with galvanised bolts. The steel flotation drums are treated with a coal tar epoxy paint, plus antifouling paint for protection. It is expected that a commercial unit would use ferrocement pontoons.

Initially a coir rope was used as the settling material but it was found that the coir rotted very quickly. Unfortunately synthetic ropes did not provide a good enough settlement, so now ropes of coir are blended with polypropylene or polyethylene. It is hoped that this combined rope will save the need for thinning the mussels which is an expensive business. These ropes, up to 4 m long, are suspended from a 3 mm steel dropper which spans the distance from the timber battens to the water level. Dr MacIntyre's results have been encouraging with growth rates of 35–55 mm a year, which are not quite as good as in Spain, but considerably better than the bouchot system in France.

Mussel Culture in New Zealand

The experimental culture of the mussel, *Perna canaliculus*, was initiated in New Zealand in 1969 (Sorensen, 1970), as a result of the Marine Farming Act of 1968 which was set up to facilitate the establishment and development of marine fish farming. Two experiments have so far been carried out on mussel farming, using the Spanish technique of floating raft culture.

Fig. 110   Buoyed frames, each carrying 28 ropes, 3 m long, for mussel culture at Linne Mhuirich in Loch Sween, Scotland.
(*Credit—P. H. Milne*)

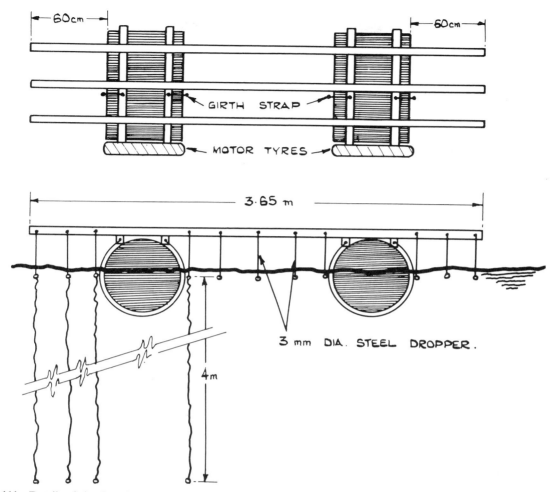

Fig. 111   Details of simple raft module for the rope culture of mussels in Australia, using timber battens and 200 litre drums.   (*Credit—by courtesy of Dr R. J. MacIntyre*)

The first location was near Coromandel, where a pontoon of oil drums was used for spat collection and culture. The spat here grew quickly and attained a length of over 10 cm in one year. The second location was in Keneperu Sound and was established by the Victoria University of Wellington. Concrete rafts were used on this site and lengths of sisal rope were suspended for the settling and growth of spat. The results of the work in Keneperu Sound were moderately successful, but some problems were encountered which require further investigation before mussel farming can be recommended (Anon., 1971e). The Marine Department and the Fishing Industry Board are now co-operating with Victoria University to undertake studies on the farming of mussels in New Zealand.

Scallop Culture in Japan

Although the Japanese scallop, *Patinopecten yessoensis*, is not yet cultivated commercially it is of potential economic value to Japan. It is one of the shellfish selected for culture studies by Professor Takeo Imai at his Oyster Research Institute at Mohne Inlet, Kesennuma, Japan (Ryther and Bardach, 1968).

At Mohne Inlet 180 1,000 litre polyethylene tanks are suspended from a raft in the sea, and seawater is pumped ashore for filtering before being fed to the tanks. The young larvae are also fed cultures of unicellular algae until they have reached the juvenile stage and are ready for setting. The young seed scallops are then suspended in the sea in a 0·3 × 1 m rectangular metal frame and held in place between two layers of loose fibre

Fig. 112 Experimental mussel raft moored in Sydney Harbour in Australia. (*Credit—by courtesy of World Fishing*)

filling and nylon mesh netting. As the scallops grow the fibre filling is removed from the sandwich frames. This method is, however, expensive and alternative tests have also been carried out attaching the scallops to ropes as in the raft culture of oysters. Both these techniques result in a rapid growth which produces a high quality of shellfish in two years, about half the time for natural growth on the seabed. The commercial success of scallop farming in Japan will depend on the development of economical methods for scallop fattening due to the present high cost of maintaining such small sandwich frames. It is possible that the use of Netlon bags (Nortene, 1969) might reduce these costs in the future.

(III) FISH

Yellow-tail Culture in Japan

The most popular method of cultivating yellow-tail, *Seriola quinqueradiata*, in Japan is to use floating net cages, moored in the sheltered bays of the south-west coast, especially in the Inland Sea. The most common is a single net cage joined together to form a group of 4–10, but others exist with double layers of net.

Single net cages are simple affairs constructed out of bamboo and floats, originally oil drums, but these are now being slowly replaced with styrofoam cylinders. A bamboo framework with walk-

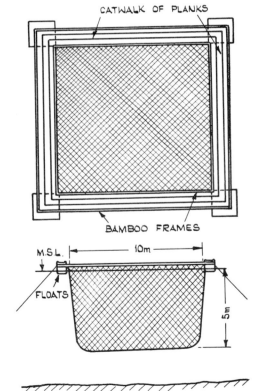

Fig. 113 Details of floating net cage with buoyancy collar for farming yellow-tail in Japan. (*Credit—from Allen, 1968*)

138

way is constructed to give a 10 m square central section. Bag nets of synthetic mesh, 10 m square and 3–5 m deep are then suspended from the floating collar, and the bottom corners of the net weighted to keep its shape, Fig. 113 (Allen, 1968). Often up to 10 of these bag nets are joined together to form a large raft, with anchors at each corner.

The double layer nets are more complex structures, and are only used in areas where net damage due to predators or wave action is possible. These nets are hung either side of the floating walkway collar, as shown in Fig. 114 (Harada, 1970). These double bag nets tend to be smaller than the single bag nets due to the extra weight of mesh and measure 7·2 m square by 3·6 m deep.

The benefit of the floating net cage type of cultivation is that the cages can be moved easily in winter into more sheltered locations, especially for maintenance. This system has been adopted for a crofter type of fish farming and the floating net raft shown in Fig. 115, belonged to a small village near Tokyo. This style of fish cultivation has increased the sea area available for fish farming, and the net cage method is progressively being used in deeper water. In offshore locations exposed to wave action, floating boom breakwaters are often used to give protection to the net cages.

Until now all the yellow-tail have been cultivated using trash fish as food, but as this is now becoming scarce, alternative sources of food, such as soya beans, fish meal, or yeast cultivated on waste petroleum or wood pulp are now being tested (Hempel, 1970).

At present the fish generally reach marketable size by the autumn when there is a glut on the

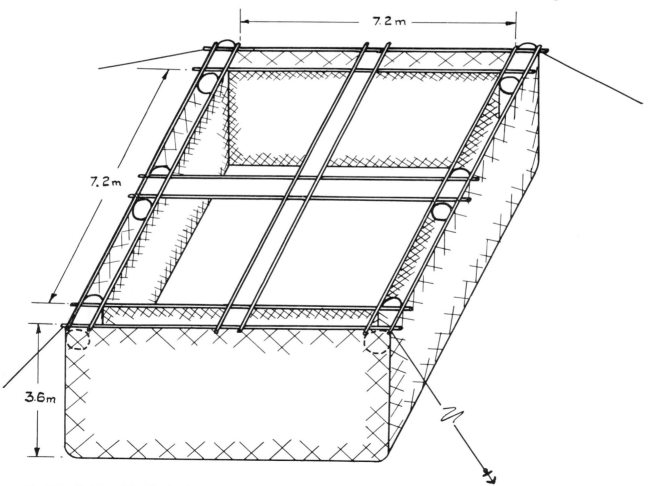

Fig. 114 Details of double floating net cage with bamboo framework. Outer predator net protects fine inner mesh in exposed situations. (*Credit—from Harada, 1970*)

market with a seasonal fall in prices. Unfortunately the normal winter water temperatures are too low without fish mortalities, so some of the operators are now moving their net cages to nearby coastal power stations to take advantage of the warmer waters which allow a longer harvesting period.

Rainbow Trout Culture in Scotland

As discussed earlier, in Chapter 6, Marine Harvest Limited established a rainbow trout, *Salmo gairdnerii* farm at Lochailort in 1966. Once the fish are acclimatised to sea water they are transferred to floating cages in the sea. The development of the engineering techniques used for these cages illustrates some of the problems of floating culture techniques. The original cages were constructed with a timber frame, 2 m cube, with a bag net of synthetic fish netting. However, these were extremely difficult to maintain and service. The second cages of buoyed tubular steel had a hexagonal collar with 2 m deep supports at the corners, and a bag net was

again hung from this framework. Once more difficulties of maintenance, feeding and mooring occurred. The third series of cages employed inflatable neoprene floats using rolled expanded metal mesh for the cage. These cages were an improvement but unfortunately during the winter the neoprene could not withstand the impact of floating ice at the top of Loch Ailort and the cages sank to the bottom with loss of buoyancy.

The first main improvement came in 1969 when polythene drums filled with polystyrene foam were used to form the collar for the hexagonal cages. These cages were again made from rolled expanded metal mesh, 2·4 m across at the top and 2 m deep, and were provided with a catwalk on top of the drums for access, Fig. 116. The success of these cages has prompted the design of larger decagonal cages, 7·5 m across at the top and 3 m deep, Fig. 117, using the same technique as the 2·4 m cages. Fig. 17 shows the floating cages at Lochailort with the hatchery and shore tanks in the background.

Fig. 115   Floating net raft for crofter type of fish farming in Japan. This raft belonged to a small village near Tokyo.
(*Credit—J. D. M. Gordon*)

Fig. 116 Small 2·4 m hexagonal fish cages for rainbow trout and salmon farming at Loch Ailort in Scotland. (*Credit—by courtesy of Marine Harvest Ltd*)

Fig. 117 Both small 2·4 m hexagonal and large 7·5 m decagonal fish cages for rainbow trout and salmon farming at Loch Ailort, connected to a spider framework with central platform. (*Credit—by courtesy of Marine Harvest Ltd*)

Access to the cages from the hatchery and food-store is streamlined by the use of an amphibious vehicle to avoid double handling. Also illustrated are the spider type frameworks employed at Loch Ailort, where up to six cages can be moored together. This system simplified first the mooring of a large number of cages in confined waters and second the access and feeding of the fish from a central platform.

The construction methods for such large units are of interest, as to their application elsewhere. As there are no convenient dockyards or slipways in the vicinity of Loch Ailort for the construction of these units, they were designed to be delivered by road and assembled on the shore at the low water spring tide mark, so that the rising tide (tidal range, 4·5 m), would float off the unit for towing to its mooring. The same technique is applied at harvest time when the cages are towed inshore at high water and moored so that they dry out at low water and the fish can be easily collected. Before towing back on location the expanded metal mesh is treated chemically to remove the attached marine fouling growths.

Rainbow Trout Culture in Australia

In Chapter 6 mention was made of the recent establishment of a rainbow trout farm at Bridport in Tasmania. In addition to the shore facilities at Bridport (Purves, 1968), the potential use of floating net cages is now being examined, Fig. 118. As the estuary at Bridport is rather exposed for floating units, it has been necessary to transfer the fish 60 km to the nearest sheltered water which has sufficient depth and flow of water for fish cages. A certain amount of success has already been achieved with these floating units (Purves, Personal communication) and it is anticipated that more cages will be installed in the near future.

Salmon Culture in Norway

In addition to the sublittoral enclosures described in Chapter 8, floating cages are also used in Norway for the farming of the Atlantic salmon, *Salmo salar*. The Brothers Grøntvedt have designed large octagonal cages 12 m across and 4 m deep, and now have 13 cages moored in coastal waters. To give protection from waves, the cages are moored behind offshore islands. It is believed that between 25–30 tonnes of salmon were sold to the market in 1971.

Salmon Culture in United States of America

As discussed in Chapter 6, hatchery techniques for the raising of Pacific salmon, *Oncorhynchus* spp.,

Fig. 118   Experimental floating cage for rainbow trout farming in Australia. (*Credit—by courtesy of Sevrup Fisheries Pty Ltd.*)

Fig. 119 Large floating fish enclosure measuring 50 m × 12 m at Reservation Bay, Washington. Enclosure is split into a small holding area and a larger test area. (*Credit—C. J. Hunter*)

are being developed. Simultaneous research has also been carried out by the Bureau of Commercial Fisheries in Washington into the use of floating sea water enclosures, rather than ponds and enclosures as discussed in Chapter 7.

The first experiments were carried out with coho salmon, *O. kisutch*, in small floating sea water pens within southern Puget Sound near Manchester. The first stock of salmon were started from fertilised eggs and raised in fresh water to smolt size. The fish were then transferred at a weight of 10–20 g each to sea water in July 1969. By feeding the fish on Oregon moist pellets, they reached a marketable size of 300–400 g within seven months (Joyner, 1970).

Subsequent experiments were carried out from a large floating structure at Reservation Bay, near Anacortes, Fig. 119. This floating enclosure measured 50 × 12 m, and comprised two separate areas, each fitted with a net enclosure 3 m deep (Hunter and Farr, 1970). The platforms round the enclosure were constructed from polystyrene covered by plywood in modules of 1·2 m × 2·4 m × 0·6 m. These units were connected together in sections using continuous 12·5 mm steel cables with

a wood spacer (10 × 10 cm) between each unit, Fig. 120. The units were assembled into sections in the intertidal zone at low water, and floated into place on the rising tide. Flotation of the structure was excellent with only 15 cm of the floats submerged, Fig. 121 and even 1 m high waves did not suspend operations. The bag nets used were 15-thread 7·5 cm stretched mesh nylon hung from the platform.

These structures were designed carefully to withstand a considerable amount of wave and wind action and have been a complete success having now withstood three seasons in some very stormy weather.

Plaice Culture in Britain

Due to the increased water interchange in net cages, these have found favour with fish farmers in Japan as the fish can be stocked at much higher densities than in either intertidal or sublittoral enclosures.

The White Fish Authority recognised this development and initiated cage trials in 1969 for the cultivation of plaice.

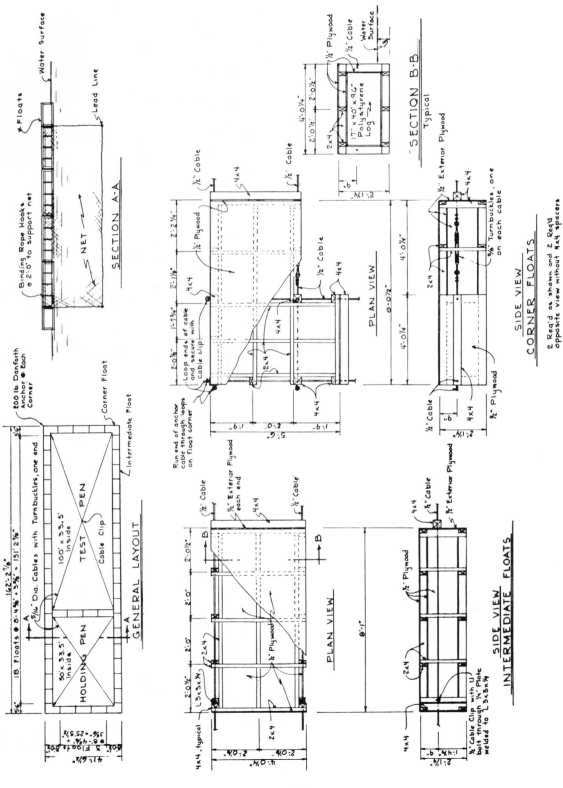

Fig. 120   Details of large floating fish enclosure at Reservation Bay, Washington.

(*Credit—from Hunter and Farr, 1970 by courtesy of the editor of J. Fish. Res. Bd. Canada*)

Fig. 121   Gill nets set inside test area of large floating fish enclosure at Reservation Bay, Washington. (*Credit – C. J. Hunter*)

The first cages were very small, 2 m square by 1 m deep, Fig. 122, and were moored adjacent to the Fish Cultivation Unit at Ardtoe, Argyll in Scotland. Subsequent experiments have been conducted in various shapes of cages, square, rectangular, circular and hexagonal.

The White Fish Authority estimate that in some of the experimental cages, measuring $6 \times 3 \times 1$ m, it is possible to grow plaice to 25 cm in 24 months from hatching, and at densities giving a production of 1 tonne every two years (Richardson, 1971). It appears, therefore, that cages of this size would be eminently suitable for fish farming in small crofting communities, probably working within a co-operative organisation as in Japan.

Tuna Culture in Japan

An investigation into the possibility of bluefin tuna culture was started in Japan in 1971, and is being carried out by the Shizuoka Prefectural Fisheries experimental station (Anon., 1971g). The fish were caught initially by setnets in Suruga Bay and trans-

Fig. 122   Experimental floating fish cage, 2 m square by 1 m deep, for plaice culture in Britain. (*Credit—Michael Wood, by courtesy of the White Fish Authority*)

ferred to two octagonal floating pens in Uchiura Bay, off Numazu City, at a density of 100 fish per enclosure. On transfer the fish weighed 300–450 g, and it is hoped after one year they will have increased in weight to 2–8 kg. One problem is expected during the winter months, and that is temperature, since normally the fish live in deep warmer water than the surface coastal colder waters at Uchiura Bay. It is also hoped that the bluefins will eventually spawn in captivity.

4
PROBLEMS

Chapter 11
Effect of Marine Enclosures on Environment

It is not only necessary to carry out a preliminary hydrographic survey of a site to determine its suitability for providing a marine environment (as discussed in Chapter 4), but the effect of any structures erected for the retention of the farmed species must also be assessed since it is probable that these may modify the original hydrographic conditions. Failure to appreciate the effect of such structures on the hydrography has led to many early mistakes in the design of marine enclosures. Recently considerable hydrographic research has been carried out in marine enclosures, in (a) intertidal, (b) sublittoral, (c) seabed, and (d) floating, and as a result it is now possible to appreciate some of the environmental problems before construction.

(a) INTERTIDAL

Ardtoe

The first marine enclosure to be constructed in Britain was built for the White Fish Authority at Ardtoe, Argyll in Scotland. The original pond covered 2 ha and was designed in the intertidal zone for research into the cultivation and raising of plaice, as described in Chapter 7. A plan of the pond is shown on Fig. 123(a) with the intertidal beach contours.

Hydrographic research in the area (Milne, 1972d) which has a 4·25 m tidal range, had indicated salinities believed to be acceptable for a marine fish farm, but as a small stream flowed through the site, a diversion pipe was provided to help reduce the fresh water inflow. Sluice valves and weirs were installed for the admission and discharge of sea water as already described. As discussed in Chapter 4, the ideal operational pond level is at the high water neap tide mark. At Ardtoe, however, a level between high water neaps and springs was chosen, with a view to sea water interchange six days per

fortnight in a similar manner to the mullet ponds at Audenge already described in Chapter 7.

Hydrography, carried out after the construction of the pond in the Autumn of 1965, showed that the sluices were adequate to empty and fill the pond as specified (Milne, 1971c). Using phosphate as a tracer, water movement could be detected even at the south-east end of the enclosure, furthest from the sluices (both sets open during the experiment), as shown on Fig. 123(b). The fresh water surface sluices, Fig. 65, had been designed to skim off the fresh water from the surface layers, in situations such as shown in the salinity profile of Fig. 123(c) before replenishment. Three days later after the admission of sea water through the bottom sluices, there was an overall rise in salinity, as shown in the profile on Fig. 123(d), but unfortunately, westerly winds during this period tended to blow the remaining less saline water into the south-east corner, where there was no method of discharge. After these experiments took place, a combination of heavy rainfall during neap tides reduced the pond's salinity when the tides were not suitable for replenishment. Even at the next set of spring tides the salinity gradient into the south-east corner, set up by north-westerly winds, such as is shown on Fig. 123(d), made it impossible to provide a satisfactory interchange of sea water.

To prevent a recurrence of these problems of inundation by fresh water in the south-east corner, it was concluded, after a hydrological survey, that the solution to the problem was to divert as much as possible of the fresh water run-off from the surrounding catchment. This was effected by the construction of a rockfill dam, Figs. 67–69 in the south-east corner, as shown on Fig. 124(a). This created a fresh water pond and a pipeline was installed to divert the fresh water out to sea. This new dam reduced the salt water pond area to 1·2 ha.

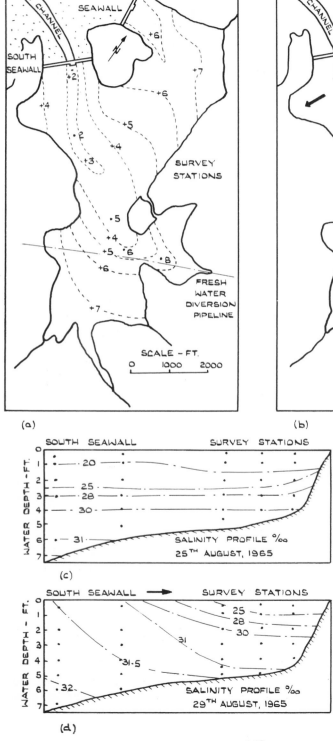

Fig. 123 Hydrographic research at Ardtoe in 1965:

(a) Plan of Ardtoe fish enclosure in August 1965, showing seabed contours and survey stations on centre line profile.

(b) Plan of Ardtoe fish enclosure in August 1965, showing water movement within the enclosure with both sets of sluices open.

(c) Salinity profile at Ardtoe showing stratification prior to replenishment, 26th August 1965.

(d) Salinity profile at Ardtoe, showing salinity gradient due to north-westerly wind blowing, 29th August 1965.

(Credit—from Milne 1971c)

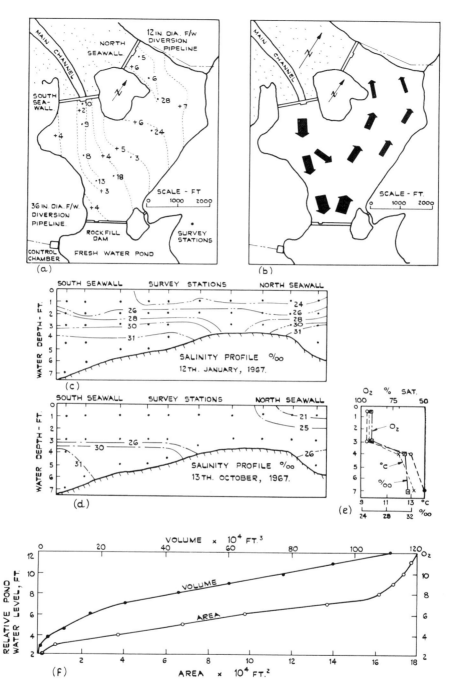

Fig. 124 Hydrographic research at Ardtoe from 1966–67:
(a) Plan of Ardtoe fish enclosure in August 1966, showing seabed contours and survey stations on centre line profile.
(b) Plan of Ardtoe fish enclosure in August 1966, showing water movement within the enclosure with south seawall sluices open.
(c) Salinity profile at Ardtoe, showing effect of using both north and south seawall sluices for replenishment, 12th January 1967.
(d) Salinity profile at Ardtoe, showing effect of using only south seawall sluices for replenishment, 13th October 1967.
(e) Vertical temperature, salinity and dissolved oxygen profile at the south seawall, 13th October 1967.
(f) Volume/area curve for Ardtoe enclosure relative to pond water level.

(*Credit—from Milne, 1971c*)

Subsequent hydrographic investigations carried out at Ardtoe (Milne, 1971c), studied the best methods of operating the seawall sluice valves to achieve the maximum sea water interchange. Early experiments with both sets of sluice valves open on the rising tide produced a two pronged salinity wedge advance as shown on the centre line profile of Fig. 124(c). However, less saline water tended to collect in the middle and north-east corners. To effectively remove this water, it was recommended that, prior to replenishment, only the north sea-wall surface weir was used, and that inflow should be regulated through the south seawall sluices. The results of one such hydrographic experiment are shown in Fig. 124(b), where, again, phosphate was used as a tracer to study the water movement. After the less saline water is removed, both seawall sluices may be used for replenishment. Operation of the sluices in this manner produces a salinity profile similar to that shown in Fig. 124(d). On this occasion only a small amount of replenishment was effected, bringing into the pond a layer of denser, warmer water. In the absence of wind to create mixing, this produced a thermocline as shown on Fig. 124(e), with a lowering of the dissolved oxygen content. Such a situation could be critical if it existed throughout a fish enclosure, but on this occasion it only occurred at the south seawall, Fig. 124(d).

For the successful management of an intertidal enclosure, it is essential to appreciate the variation in the pond area and volume with water level, Fig. 124(f), to ensure that effective interchange is carried out when the tides are suitable. If the selected pond level lies below the high water neap tide mark, interchange can be effected at any time, provided the sluices are designed adequately.

A study of the interchange at Ardtoe shows that a figure of 30% in 24 hours is normal at springs, but figures of 35 and 41% have been recorded during experimental work. These low figures for Ardtoe are due to the same channel of Sailean Dubh being used for the admission of sea water and the discharge of less saline water, so that only every second high water can be used for replenishment. If the pond at Ardtoe was to be operated at the high water neap tide level, Fig. 13, allowing interchange every tide, the figures for replenishment would be 60% at neaps and 120% at springs (Milne, 1971c). The diversion of the fresh water at Ardtoe has solved the problem of inundation by fresh water, and the pond is now fully operational again with a marine environment.

In the design of an intertidal pond it is essential to decide at the outset whether it is to be operated with a minimum draw-down level, or if the pond is to be drained frequently. In the first case, after immersion, the natural fauna is subjected to a change from an intertidal to a sublittoral regime, and this can create problems with dead and dying weed if it is not cleared initially. Once the pond is flooded there will be a natural influx of colonising species, which, if given a submerged habitat will produce a sublittoral fauna. This may be advantageous if it assists in producing extra food for the farmed species. However, if the pond is drained frequently for either harvesting or maintenance the situation described above will not occur.

(b) SUBLITTORAL

Adoike

As mentioned in Chapter 8, Adoike is the largest sublittoral pond enclosure in Japan, and is used for the cultivation of yellow-tail. As shown in Figs. 73, 74, the bay at Adoike was enclosed by the construction of a sea embankment. The average depth of the pond is 8 m and the maximum 12 m. Two sets of sluices were provided in the embankment for water circulation (Tamura and Yamada, 1963). At A four large sluice gates, 3 m deep, span a gap of 11·55 m. A gantry is provided as shown on Figs. 75–76 for the replacement and removal of the mesh screens for cleaning. At B the gap is 2·60 m and 9·50 m wide and is filled with large diameter pipes laid horizontally to admit sea water with a mesh net on the inside for screening, Fig. 75. These sluices are open at all times, so the pond is tidal with a range of 1·5–1·8 m.

According to a hydrographic survey carried out in September 1962 by the Agriculture Faculty of Kagawa University (Tamura and Yamada, 1963), the interchange of water in the enclosure is no more than 30% in a 24 hour period. The observers found that the replenishment appeared to occur in the vicinity of the sluice gates, and this was where the fish were seen circling. Very few fish were seen in the inner parts of the enclosure where there was a reduction in the oxygen content. Tamura and Yamada concluded that unfortunately the usage of the enclosure was restricted to one-third of the area as a result of insufficient research work on the interchangeability of water. As will be seen from Fig. 73, the sluice gates A and B are very close together, and it would have been better if the sluices had been at either end of the embankment. Alternatively, mechanical means of aeration in the

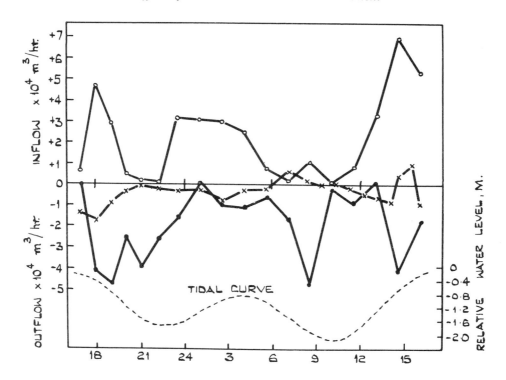

Fig. 125 Diurnal variation of water-exchange in the yellowtail farm at Hitsuishi in Japan. Plan of area shown in Fig. 77.

Legend: ○ inflow through net barrier, m³/hr
        ● outflow through net barrier, m³/hr
        × inflow and outflow through sluice gate, m³/hr

(*Credit—from Inoue* et al., *1966*)

innermost parts would overcome the dissolved oxygen problem.

Hitsuishi

The area chosen for a yellow-tail farm at Hitsuishi lay between two islands as described in Chapter 8. This is a very shallow sublittoral fish farm with an average depth of 3 m and a maximum depth of 5 m as shown on Fig. 77. Due to the shallowness of the basin at low water the tidal water exchange plays a leading role in the dissolved oxygen budget, and thus determines the productivity and optimum number of fish which can be cultivated successfully (Inoue *et al.*, 1966).

To determine the optimum stocking density at Hitsuishi, Inoue *et al.* of the Agriculture Faculty of Kagawa University carried out a hydrographic investigation during a neap tide, since this is the time of smallest water exchange and low dissolved oxygen content. In fact an oxygen deficit has been commonly experienced in some farms at neaps and mechanical methods of aeration have had to be employed to overcome this problem.

Although a gate was designed for the southern embankment it was found that there was little water flowing in or out at neap tides. However, the water exchange through the 350 m long net barrier at the northern end was considerable, Fig. 125, with current velocities of 5–10 cm/sec (Inoue *et al.*, 1966). The large water exchange is due to the difference in volume of $8 \cdot 7 \times 10^4$ m³ at low water and $22 \cdot 3 \times 10^4$ m³ at high water.

On the basis of the small water exchange rate at neap tides Inoue (1965) considers that the maximum number of yellow-tail at 300 g in August should be restricted to $6 \cdot 5 \times 10^4$ for extra safety but could be raised in auspicious circumstances to

$15 \cdot 5 \times 10^4$. When the farm was stocked at the higher limit in 1962 the average weight of the fish harvested at the end of November was 825 g. However, by reducing the stocking density in successive years the average weight of the harvested fish rose to 1,150 g in 1964.

Ieshima

This yellow-tail farm in Japan covers an area of 91·7 hectares, and as discussed in Chapter 8, is an open bay enclosed by a 400 m long net barrier, Fig. 80. To investigate the environmental fluctuations within Ieshima fish farm and the effect of the net barrier a hydrographic survey was carried out both inside and outside the farm on 26th August 1964 (Sugimoto *et al.*, 1966).

Ieshima is a relatively deep fish farm, the main area having depths of 4–10 m, whilst the net barrier is 15 m deep in the centre. A study of the water currents indicated inflow at the southern end of the net, an anti-clockwise motion in the fish farm with outflow at the northern end. A study of the actual current measurements recorded showed a decrease of 46% from outside the net to the fish farm with an

exchange flow ratio of 1·31 per tide. This considerable speed reduction was considered by Sugimoto *et al.* to lie in the resistance of the net across the bay.

A study was also made of samples of the seabed for sulphides, and samples taken round the feeding stations showed considerable contamination of the bottom. Since the farm had only been in operation for two years, this contamination due to feeding the yellow-tail with trash fish was considered high by Sugimoto *et al.* who recommended that attention be paid to methods of preventing the successive contamination of the bottom with the passage of time. Due to the small exchange flow ratio the stocking in 1964 was near the maximum at 1 kg/3·8 m³.

Megishima

The 5·6 ha yellow-tail enclosure at Megishima, described in Chapter 8, Fig. 82, was also studied by Sugimoto *et al.* (1966) who carried out an internal and external hydrographic survey of the fish farm on 22nd August 1964. This again is a comparatively shallow fish farm with the main area having depths of 2–5 m. With a 2 m tidal range the

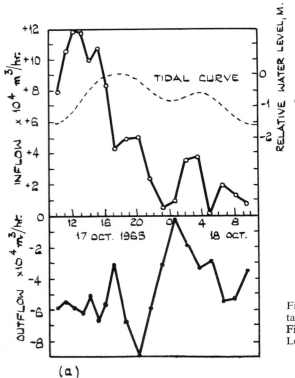

(a)

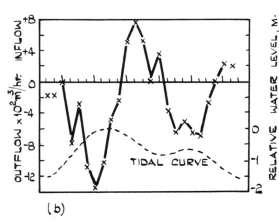

(b)

Fig. 126 Diurnal variation of water-exchange in the yellow-tail farm at Tanoura in Syozu-sima. Plan of area shown in Fig. 83.
Legend: (a) ○ inflow through net barrier, m³/hr
          ● outflow through a net barrier, m³/hr
    (b) × inflow and outflow through channel, m³/hr
                  (*Credit—from Inoue* et al., *1970*)

interchange in these enclosures is relatively high with an exchange flow ratio of 10·3 per tide. On the flood tide the currents run from west to east and vice versa on the ebb tide.

The effect of the rubble mound breakwaters and the net barrier on the current speeds is quite considerable, with a reduction inside the fish farm to 25% of the current speeds outside. The maximum current speed recorded inside was 12 cm/s. Bottom samples were also analysed and showed a finer particle size inside to outside the fish farm. Sulphide amounts in the bottom samples were higher inside the fish farm than outside and indicated considerable contamination of the bottom due to waste food and faecal material over the five year period of operation of the fish farm. The turbidity of the water was also greater inside the fish farm. Only the large exchange flow ratio per tide enables this enclosure to be stocked at a density approaching 1 kg/0·57 m³.

Tanoura

The yellow-tail fish farm at Tanoura in Japan, described in Chapter 8, is similar to Ieshima as it is a large bay sealed by a 560 m long net barrier, Fig. 83. A hydrographic survey of Tanoura was carried out by Inoue *et al.* (1970) on 17th and 18th October 1965 to determine the environmental influence of the net barrier five years after the original construction in 1960.

In addition to the sea water exchange through the net barrier a small artificial channel 1·8 m wide and 150 m long was constructed through the narrow neck of land separating the innermost part of the bay with the open sea, Fig. 83. This small channel, however, has very little influence on the sea water exchange in the fish farm as it only amounts to 0·22% on the flood tide and 0·7% on the ebb tide. Inoue *et al.* carried out their studies over a neap tide, when the water exchange was least and measured the inflow and outflow through both the net barrier and the artificial channel, Fig. 126(a) and (b). In general the sea water flowed into the fish farm in a surface layer on the flood tide, and on the ebb tide flowed out through the western half of the net at its deepest section, Fig. 83, with little exchange through the eastern half of the net. On the basis of water exchange at neap tides, Inoue (1965), estimates that the optimum stocking density in late August with 400 g yellow-tail should be 40 × 10⁴ fish.

It is however extremely interesting to study the production record for this fish farm in the period 1960–65, Fig. 127, and compare the average size of the harvested fish with the stocking density.

NUMBER OF FISH CULTURED IN TANOURA FISH FARM
(Inoue *et al.*, 1970)

Year	1960	1961	1962	1963	1964	1965
Number of fish stocked × 10⁴	16·8	37·6	30·2	18·2	37·3	46·0
Number of fish harvested × 10⁴	12·9	16·6	28·3	17·1	36·4	42·0
Average body weight in December, kg	1·39	1·19	1·07	1·15	1·13	0·79

Fig. 127   Number of fish cultured in Tanoura fish farm from 1960–1965. (*Credit—from Inoue* et al., *1970*)

In Fig. 127 it will immediately be seen that the average weight of the harvested fish is directly proportional to the stocking density indicating that the water exchange and dissolved oxygen requirements of the site limits the optimum production. A farmer therefore has to decide on the required weight of fish he wishes for harvesting before deciding on his stocking density because of the direct relationship.

Strom Loch

This rainbow trout farm was constructed in Scotland in 1968 as described in Chapter 8, Fig. 84. The farm lies at right angles to the shore, Fig. 128(a) and the water flow is provided by the variation in tidal level. The spring tide range for the Shetlands is 1·5 m, but with restrictions at the entrance to Stromness Voe and Strom Loch the tidal variation is reduced to 0·3 m at the fish farm. With an offset entrance channel at Strom Loch, the tidal flow on both the flood and ebb streams is from north to south through the fish farm.

A hydrographic survey at the fish farm on 25th August 1970, shown as a salinity and temperature profile on Fig. 128(b), indicated only slight variations (Milne, 1971c). The high temperatures and low salinities recorded in Strom Loch in midsummer are a result of its being isolated from the continental shelf area by two narrow, shallow sills as shown on Fig. 128(c). At the entrance to Stromness Voe the seabed falls to 40 metres within a short distance giving oceanic water, as shown in Fig. 128(d) for salinity. Both Stromness Voe and Strom Loch are therefore typical Scandinavian fiordic sea lochs, which with tidal ranges less than 0·5 m tend to be stagnant. From the entrance of

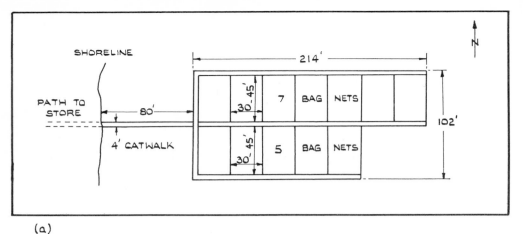

(a)

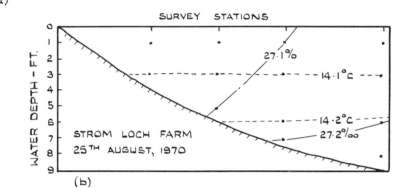

(b)

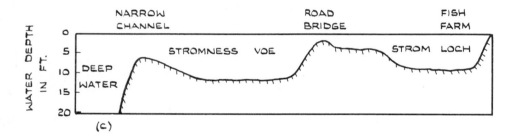

(c)

WATER DEPTH FT.	STROMNESS VOE ENTRANCE				STROM LOCH ENTRANCE				FISH FARM ENCLOSURES			
	OUTSIDE		INSIDE		OUTSIDE		INSIDE		INSHORE		OFFSHORE	
	SAL.‰	TEMP.°C	SAL.‰	TEMP.°C	SAL.‰	TEMP.°C	SAL.‰	TEMP.°C	SAL.‰	TEMP.°C	SAL.‰	TEMP.°C
0	34.95	11.6	34.95	11.8	34.50	12.4	33.71	12.8	27.05	14.0	27.15	14.0
3	35.05	11.6	35.05	11.6	34.90	12.3	—	—	27.1	14.1	27.15	14.1
6	35.15	11.6	—	—	—	—	—	—	27.15	14.2	27.2	14.2
9	—	—	—	—	—	—	—	—	—	—	27.25	14.2

(d)

Fig. 128 Hydrographic research at Strom Loch Fish Farm in Shetland:
(a) Plan of Strom Loch Fish Farm.
(b) Salinity and temperature profile, 25th August 1970.
(c) Seabed profile of Stromness Voe and Strom Loch.

(d) Salinity and temperature observations from the entrance of Stromness Voe to Strom Loch Fish Farm.

(*Credit—from Milne, 1971c*)

154

Stromness Voe to the fish farm, a distance of 6 km, the salinity drops from 35 to 27‰ and the temperature rises from 11·6 to 14·2°C, two considerable variations, Fig. 128(d).

As described in Chapter 8, the rainbow trout at Strom Loch were kept in bag nets suspended from a fixed framework, Fig. 86. From the hydrographic observations, it was estimated that there were 1½ water changes per hour at neaps and three water changes per hour at springs in each bag net. For rainbow trout, the maximum stocking capacity of each bag net, assuming 100% dissolved oxygen saturated sea water was therefore 10,000 225 g fish, which would provide a minimum of exchange at neaps, and double that at springs. However, when the author visited the farm in August 1970 to carry out a hydrographic survey he found the farm only stocked to 40% of its capacity due to lack of oxygen. The above stocking calculations rely on a supply of 100% dissolved oxygen saturated sea water, whereas as shown in Fig. 128(d) and mentioned earlier, Strom Loch is a typically stagnant fiordic sea loch.

These assumptions were confirmed on two SCUBA dives in Strom Loch (Milne, 1971c), when large anaerobic areas with patches of waste food, faecal material, and dead crabs were observed in the vicinity of the fish farm. At this fish farm the rainbow trout were fed trash fish and this build-up of bottom contamination had occurred in only two years. These observations are similar to those of Inoue *et al.* (1966) and Sugimoto *et al.* (1966) in Japanese fish farms, discussed earlier in this chapter, and it is obvious that attention must be paid to keep the seabed free of contamination when the fish are fed with trash fish.

Although Strom Loch provides a very sheltered location with minimal water currents and water forces on the netting, these are at variance with the requirements of a highly oxygenated water supply, either from wind action or tidal current turbulence. If, however, mechanical methods of aeration were to be used these enclosures could be stocked to their full capacity, provided consideration is given to the seabed contamination.

Faery Isles

These two 12 metre square fish enclosures were constructed in 1969 at Faery Isles, Loch Sween, Argyll, Figs. 102, 103, for research into the civil engineering aspects of sublittoral enclosures, as discussed in Chapter 8. The location of the enclosures at Faery Isles is shown in Fig. 129(a).

Prior to construction, regular hydrographical and meteorological observations were recorded (Milne, 1972d). These observations commenced in 1966 and ensured that variations in temperature, salinity, dissolved oxygen content and current velocities were known so that water movement could be assessed. As the area has a small tidal range for British waters, 1·5 m, the water tends to be warmer in the summer than nearby coastal waters, and hence marine fouling exposure tests were carried out from a raft at Faery Isles, for the selection of the fish mesh retention material, as discussed in Chapter 5. In this case, galvanised weldmesh proved to be the material least susceptible to marine fouling.

Since the construction of the enclosures in April 1969 the author has carried out regular hydrographic and current observations in the vicinity of the enclosures to assess their influence on the marine environment (Milne, 1971c). Initial observations indicated a slowing down of surface currents, and this was due to drifting seaweed catching on the netting between the high and low water marks, thereby impeding the water flow, so surface drift weed has been cleared from time to time. The underwater section had not, however, been cleared during the observations to be discussed below.

It would appear that the selection of galvanised weldmesh is a good choice for emplacement, rigidity, and marine fouling, provided it is cleaned once a year (Milne, 1970b). From the discussion in Chapter 5 it appears that galvanised weldmesh is also superseding synthetic mesh in Japan as well.

The effect of not cleaning the mesh at Faery Isles is illustrated in temperature and salinity profiles through the enclosures on 14th and 15th April 1971, Figs. 129(b), (c) and (d), exactly two years after construction (Milne, 1971c). Fig. 129(b) shows a salinity profile at Faery Isles and indicates the arresting of the 32·0‰ isohaline at the fish enclosures. Fig. 129(c) shows the conditions on a warm and sunny afternoon while Fig. 129(d) shows the conditions after overnight frost. Both sets of observations were taken at high water spring tides and the location of the mesh structure is shown by dotted lines. The vertical lines represent the 25 mm mesh and the inclined line the 75 mm mesh on the offshore face as a predator and trash net, Fig. 23, as discussed in Chapter 5.

The indications of Figs. 129(c) and (d) are that there is less water movement in the vicinity of the offshore face of the structure, especially under the catwalk between the 75 and 25 mm mesh nets.

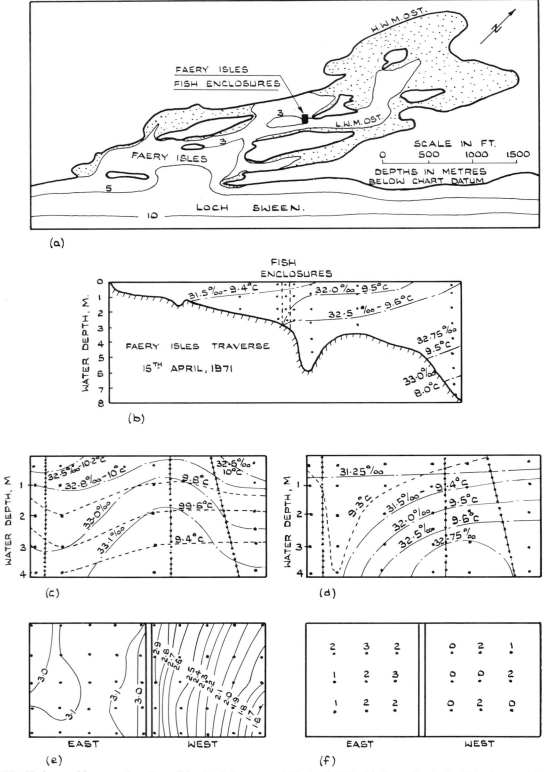

Fig. 129 Hydrographic research at Faery Isles, Loch Sween: (a) Location of Faery Isles fish enclosures, at Faery Isles, Loch Sween. (b) Salinity and temperature profile of Faery Isles, 15th April 1971. (c) Salinity and temperature profile through Faery Isles fish enclosures, 14th April 1971. (d) Salinity and temperature profile through Faery Isles fish enclosures, 15th April 1971. (e) Plan of seabed within the two 12 metre square fish enclosures at Faery Isles, April 1970. Depth below chart datum in metres. (f) Deposition of seabed sediment in cm within the two 12 metre square fish enclosures at Faery Isles, from April 1970 to April 1971.

(*Credit—from Milne, 1971c*)

This area has also been shaded from the sun and frost, giving lower and higher temperature readings respectively. These observations of a slowing down of water movement have also been confirmed by current measurements within the enclosures.

The seabed sediment within the enclosure has also been studied since construction using SCUBA divers (Milne, 1971c). Initial observations during the first year showed some slight deposition at the seabed seal of the mesh netting. A precise survey of the seabed contours with a 3 m grid in April 1970 gave the contours shown in Fig. 129(e), showing the east enclosure in mid-channel. The initial location for the enclosures had straddled the channel, but they were moved to one side to allow yachtsmen and boating enthusiasts deeper water access to the upper area of Faery Isles, Fig. 129(a).

Sediment markers consisting of 2 cm dowelling rod, marked at 2 cm vertical intervals, were driven firmly into the seabed sediment in a 3 m grid, and regular observations made of any changes. The change in level in cm at the end of one year, in April 1971, are shown in Fig. 129(f), where the deposition of sediment confirms the earlier supposition of slower water movements. Similar findings have been reported in Japanese fish farms by Inoue *et al.* (1966, 1970) and Sugimoto *et al.* (1966) as discussed earlier in this chapter. The maximum deposition at Faery Isles of two points of 3 cm has occurred in the east enclosure in the deepest area. An analysis of the seabed sediment in this area gave 74% silt and clay, 20% fine sand, 5% medium sand and 1% coarse sand. For comparison, rates of sediment accumulation in modern estuaries have been estimated at 3 cm per year in the Mississippi Delta deposits, but are high compared to normal estuaries of 0·02 cm per year. These indications confirm the earlier engineering premise that the mesh netting should be cleaned at least once a year, otherwise the enclosures will become a silt trap.

(c) SEABED

Kinlochbervie

Seabed enclosures were constructed for Kinlochbervie Shellfish Company in Lochs, Inchard and Laxford in Scotland for lobster culture as described in Chapter 9. An opportunity for comparative hydrographic observations was taken in 1970 when the author was invited to inspect the 12 m × 12 m × 2·4 m enclosures on the seabed at a depth of 12 m. Although the lobsters only utilise the floor of the enclosure, the 2·4 m height was designed to allow for access by SCUBA divers.

Hydrographic observations in the enclosures compared very favourably with previous records (Milne, 1972d). By carefully selecting sites for the enclosures outwith the zone of surface turbulence, but with good water movement, no problems with regard to water exchange have arisen or are expected. Neither have any problems arisen in relation to the seabed sediment in connection with deposition or scour (Milne, 1971c).

(d) FLOATING

Floating Net Cages

The use of floating net cages for the cultivation of yellow-tail is extremely popular in Japan, as described in Chapter 10, Figs. 113–115. The major difference in the water exchange between the net barrier type of farms, discussed earlier and the net floating cage is that there is no volume change due to the tidal rise and fall. It is purely tidal currents and wind action which produce the water exchange within the net cage.

The Japanese in addition to carrying out environmental research in net barrier fish farms have also investigated the water movements in floating net cages. One such investigation was carried out by Hisaoka *et al.* (1966) at Daio Bay at Ninoshima in Hiroshima Prefecture. These studies were conducted on two occasions in June and August 1965.

Current measurements were made outside and inside Daio Bay, and also inside a floating net cage, 5 m × 5 m × 4·5 m. The current speeds recorded outside the bay area varied up to a maximum of 26 cm/s, but those inside the bay only reached a maximum of 15 cm/s. Inside the floating net cage the currents were again reduced. In June the measurements were carried out in a cage with a 5 mm clean mesh, the net only having just been immersed in the sea a few days before the investigation. The currents, inside the net cage, on this occasion were only 60·3% of those recorded outside the net cage. By August the same yellow-tail had grown from an average weight of 12 g to 209 g, and the mesh of the net cage increased accordingly to 3 cm. In August the reduction in current inside the net was only 70·2% due not only to the increase in mesh size, but to the fact that the net was clean having only been changed on that day.

During the investigation it was noticed that if the outside currents were more than 4 cm/s the currents inside were in the same direction. However, at speeds below 4 cm/s, the swimming action of the fish caused circulating currents with an eddy in the middle of the swirl.

Due to the fact that a reduction in water currents occurred with both sets of measurements which were carried out in clean nets without marine fouling, it is expected that the flow will be reduced even further in fouled nets. Not only will the flow be reduced but the net will deform more if the nets are fouled than with clean mesh (Okabayashi, 1958). This deformation in the direction of flow reduces the available volume of water for the fish, and this can be serious in a highly stocked net cage. Similarly, if there is a long period of slack water at either high or low water during neap tides the dissolved oxygen supply within the net cage could be critical. Fortunately over the tidal cycle at neaps the water volume within the bag net is changed over 200 times at Daio Bay, so this problem does not occur here, but is a point worth noting since the amount of dissolved oxygen available for fish varies with the volume of exchanged water. This high value for the water exchange ratio explains why net cages can be stocked at a very much higher density than net barrier enclosures.

Chapter 12
Control of Predators

In the marine cultivation of the sea it is essential that predatory species are controlled and removed, since they not only compete with the farmed species for the available food, but they also reduce the potential harvest. The alternative methods for control are physical and chemical.

Physical Control

For pond farming the best method of predatory fish control is to install fine mesh screens on both the inlet and outlet sluices to prevent their entry. However, it is possible that predatory larvae, of crabs for example, may pass through the screens to grow in the pond, and unless the pond is drained frequently, crab pots will be required to deal with this nuisance.

Sea birds are another problem in pond farming, especially wading birds, when the pond is shallow. If it is not possible to grade the pond sides to give a steep slope, like the shrimp and pompano ponds in Florida, Figs. 32–34, it may be necessary to provide cross-wires above the pond's surface. Alternatively bird scare noises can be operated at regular intervals to frighten off the birds.

Where the seabed is used for the laying of sessile species such as molluscs it is essential that invasions of starfish and crabs are prevented. To control possible predators on these beds it is necessary to carry out regular monitoring round the perimeter twice a week, and this can be conveniently carried out by SCUBA divers. This method is used on the oyster farms in Long Island Sound, described in Chapter 9. When a starfish invasion is detected a patrol boat equipped with a string "mop" is dragged round the periphery of the oyster bed, the starfish being caught when the string brushes across its spines (Bowbeer, 1970). These mops are cleaned every 10 minutes by dipping in very hot water from an oil-fired boiler carried on deck, Fig. 130. If a large starfish invasion is detected a 30 m diesel powered barge can be used with a powerful suction dredge to remove the predators, and on one occasion it is reported that 100 tonnes of starfish were dredged up in 3 hours in Long Island Sound. In some cases the starfish dredged up are sold as fertiliser as a side product of the oyster farming, for example in Rhode Island (Nowak, 1970). Oyster drills and borers can also be a problem, and drill traps are often the best method of control.

To prevent predation of clams by crabs and starfish, a novel method has been developed in the United States. To protect the young clams gravel or aggregate is spread to a depth of 25–75 mm over the seabed prior to planting the clams. The clams are only placed when the water temperature is above 9°C, since the clams will then burrow down under the aggregate in a shorter time, since they are more active at higher temperatures. It is claimed that this method will protect up to 80% of the stock even when crabs and starfish are present.

As mentioned before, marine organisms which compete for the food of the farmed species are considered as predators. Ascidians on mussel ropes are a problem since they not only utilise the same food chain but also cause over-crowding which restricts the development of the mussels. Where the mussels are grown in the intertidal zone ascidians are not a problem and starfish, etc., can be easily removed. To combat attack by whelks, borers and starfish, floating raft culture is often chosen as the farming method. However, since the mussel ropes are now permanently immersed, ascidians will settle and grow. The physical removal of these ascidians is quite a problem, as mentioned in Chapter 5. The best method, other than removal, to inhibit the growth of any ascidians or other organisms on these mussel ropes is to lift the ropes out of the water into air for 30–60 minutes, since this exposure kills off the sublittoral species.

Chemical Control

Although the physical methods of control discussed earlier are partially helpful in controlling predators, they are not as long lasting as the use of chemical methods. The use of chemicals and their effect on marine organisms has to be carefully investigated before they can be recommended for large scale commercial use. Fortunately Loosanoff (1960) at the Milford Laboratory of the U.S. Bureau of Commercial Fisheries has carried out many such tests to assist in predator control, mainly with shellfish.

Extreme caution should, however, be exercised when attempting to use chemical methods of control. Approval should always be requested from government fishery, health and purification bodies to determine any restrictions and regulations concerning their use.

In controlling undesirable marine predators chemicals may be used in several ways (Loosanoff, 1960):

(1) dissolved in water;

(2) spread over large areas of the seabed as a thin layer which will repel predators, such as drills, starfish, crabs, etc., or kill them or their larvae on contact;

(3) used in so-called barriers, or belts made of chemicals mixed with materials of an inert nature, such as sand or fragments of old oyster shells, and laid as a continuous band to surround the mollusc beds so that predators, such as starfish and drills, will be stopped before penetrating into the protected area;

(4) combined with shell material of living oysters, old oyster shells or spat collectors to make them either unsuitable for the existence of sponges, worms and other shell dwelling species, or to prevent their fouling with tunicates, hydroids, barnacles, *Crepidula*, worms, algae and others;

(5) combined with shells of dead oysters, or other materials used as cultch, to repel drills or to kill larvae of other undesirable species, such as flatworms, *Stylochus*, which may set on these shells and attack oyster set. A modification of the latter method may consist of using chemicals which will not only repel enemies, but simultaneously, attract oyster larvae to set in large numbers on the specially treated collectors; and

(6) incorporated as poisons in foods that will be eaten by predators.

No detailed discussion of the various chemicals that can be used for different species is given here,

Fig. 130 The regular monitoring of mollusc beds for predators is most important. In Long Island Sound when an invasion of starfish is detected a patrol boat equipped with string "mops" is dragged over the bottom to entangle the starfish. They are killed by immersion in hot water. (*Credit—United States Bureau of Commercial Fisheries*)

Fig. 131 A chemical method for controlling predators such as starfish on mollusc beds is to spray the beds twice a year with quicklime. Here a U.S. vessel pumps quicklime overboard with a distribution of 2·5 tonnes per hectare. The tank on the vessel holds 15 tonnes of quicklime. (*Credit—by courtesy of World Fishing*)

since Loosanoff (1960) in his references deals very fully with this topic. The most common method for the treatment of oyster beds is dosage with quicklime, Fig. 131, and where no suction dredge boats are available in the United States for the removal of starfish, the dosage of quicklime is approximately 2,500 kg per hectare (Bowbeer, 1970).

Such heavy dosages of quicklime or any other chemical, on entering the water become pollutants. Fortunately the organic solvents recommended by Loosanoff (1960), are virtually insoluble in water, and therefore are relatively safe to use in a marine environment. Nevertheless there is a potential danger that in the indiscriminate use of large quan-

tities of chemicals other marine species could be affected. Therefore before any chemical compound is used for predator control on a large scale it should be thoroughly tested by experienced biologists in terms of its danger to marine life and of course in relation to human safety.

Even after obtaining such advice from biologists on the use of certain chemicals, Iversen (1968) advises the marine farmer to experiment with small concentrations of chemicals on a few of his stock for varying periods of time. In this manner the farmer can ascertain the best method and strength of concentration to ensure that his large scale commercial dosages do not wipe out the farmed species as well.

Chapter 13
Marine Pollution

Throughout the chapters of this book there have been references to natural oyster and mussel fisheries in intertidal estuaries which have deteriorated and have been abandoned since the beginning of the twentieth century. Two striking examples are first the oyster fishery in Japan described in Chapter 7, where Matsushima Bay, near Sendai, was originally the centre of the seed oyster industry. Due to marine pollution the seed oyster trade has had to move further up the coast to the less heavily populated Ojika Peninsula. The second example is the shellfish fishery in the United States where 25 % of the total shellfish grounds have now been closed because of chemical pollution (Barton, 1970, 1971).

Not only is the sea polluted due to marine wastes in many areas, but the fresh water stream discharges are also severely contaminated. Iversen (1968), recommended that the best and most immediately promising locations for marine farming, were in fertile bays, estuaries, and intertidal zones that man has ready access to, and where the stock can be enclosed and protected. Unfortunately, today these areas of coastal water are near large populations, the pollution of the sea by detergents and human wastes is increasing.

At long last this problem of pollution of our rivers and seas is now receiving attention, and governmental control in the form of purification boards are helping to redress the balance. It will be appreciated that all types of pollution from sewage, industrial and pesticides are to be avoided in setting up a sea farm. FAO (1970b) classify pollutants into seven main groups:

> halogenated hydrocarbons, petroleum, inorganic chemicals, organic chemicals, nutrient chemicals, suspended solids, and radioactivity.

Hence the reason for marine farms being distributed in the unpopulated and undeveloped parts of even so-called developed countries.

The effect of domestic wastes and detergents in general reduces the oxygen content of the water, and can transmit pathogenic micro-organisms and may also render the water objectionable in colour and odour. These colour changes in estuarine waters reduce the light penetration with the subsequent devastation of aquatic plants, which in turn also reduce the oxygen content of the water. Industrial effluents which contain complex organic substances, such as from paper mills and chemical plants, and metal ions from plating factories, can taint the taste of fish and shellfish, and render them unfit for human consumption.

In the last chapter the possible use of chemicals to control predators was discussed. The pollution effect of these chemicals and other new chemicals poses a constant problem as to their effect on marine life. Before such chemicals can be used, for example in Britain, certain toxicity tests are required on fish (H.M.S.O. 1969) to determine the concentrations at which these chemicals are toxic (H.M.S.O., 1970). Mawdesley-Thomas (1971) has however expressed the view that these tests are carried out over too short a period, and the tests should be extended to ensure that potentially harmful ecological effects are avoided.

Nevertheless, the one industrial effluent which might prove beneficial for sea farming is the warm water discharge from coal, oil or atomic power stations. These discharges due to the industrial need for cooling can, however, be turned into a biological asset since cold water species grow faster in the warmer power station water and reach marketable size much quicker than in nature. The use of warm water for plaice and sole cultivation in Scotland, and lobster and oyster cultivation in the United States has already been mentioned in previous chapters, and is one aspect of marine cultivation that will develop increasingly in the future (Nash, 1968; Gaucher, 1970; Beauchamp et al., 1971).

The possible use of the warm water effluents of atomic power stations has to be viewed with caution, however, since aquatic organisms tend to concentrate radioactive material (Foster and

Davis, 1956). In Britain the radioactivity in surface and coastal waters is checked periodically by the Fisheries Radiobiological Laboratory of the Ministry of Agriculture, Fisheries and Food to ensure the safe disposal of radioactive waste (Mitchell, 1971).

A major pollutant in the intertidal zone is oil, either released from tankers at sea, or spilt accidentally at the oil field or shore terminal. These oil spills are unfortunately becoming more frequent, and their consequences are disastrous as for example the breaking up of the *Torrey Canyon* in 1967 off Lands End in Britain. Consequently as a result of this and several other oil spills, local authorities are better equipped to deal with oil spills now than ever before, and many companies market booms for oil spill containment (Anon., 1970b, 1971f, l), such as Fig. 132 (Gamlen, 1971). Compressed air bubble curtains can also be used and one was installed across the Helford River Estuary at the time of the *Torrey Canyon* disaster to protect the oyster beds in the inner estuary, Figs. 133, 134 (Milne, 1970a). Unless positive proof is available that a certain vessel was responsible for the oil spill that wipes out a farmer's stock it is very difficult for the farmer suffering extensive losses to obtain damages. There is some hope, however, as noted by Iversen (1968) in that an oil refinery has paid the municipalities in Rhode Island compensation for an oil spill on the beaches. If better detection of oil spill sources is achieved, as is hoped in the future, this would enable farmers who had stocks damaged by oil spills to claim compensation.

Due to the increase in pollution in traditional shellfish farming areas, methods of purifying for resale have been examined for many years (Nowak,

Fig. 133   Compressed air bubble barrier erected at entrance to Helford River Estuary to protect oyster beds from the oil after the *Torrey Canyon* disaster. (*Credit—P. H. Milne*)

1970). In the United States oysters from the polluted areas are transferred to clean water areas for a month before harvesting (Iversen, 1968). This is expensive since it means that the oysters have to be double handled and the relaying costs are considerable in addition to the losses from this procedure.

In Britain the Public Health (Shellfish) Regulations, passed in 1934 allow a local authority to permit the sale of shellfish from polluted areas, but only after they have been subjected to one of three basic forms of treatment:

  (i) sterilisation by heat;
  (ii) relaying in clean water; and
  (iii) purification in an approved plant.

To assist British shellfish growers, the Ministry of Agriculture, Fisheries and Food have carried out extensive research into (a) the factors influencing the pollution of shellfish, and (b) the methods available for treating them to produce clean shellfish (M.A.F.F., 1961, 1966, 1969).

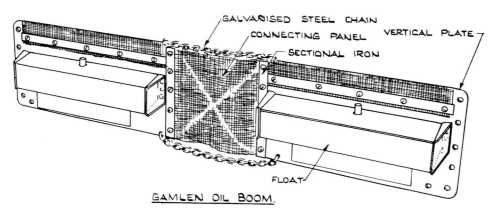

Fig. 132   Gamlen oil boom for oil spill containment.   (*Credit—by courtesy of Gamlen Chemical Co. (U.K.) Ltd*)

To save the relaying of clams in clean areas from polluted ones before harvesting, the Rhode Island Division of Fish and Game (Iversen, 1968) have recently perfected a cheaper method. This system uses ultraviolet light to sterilise the water the clams filter through their shells, and in a short time they are free from pollution. Some success with chlorinated water has also been achieved for purifying oysters and clams.

In Spain the raft culture of mussels has developed into a major industry. Although there is no industrial pollution in the Spanish rias there is some biological pollution, which makes cleansing desirable for 100% purity. To carry out the purification of these mussels a central cleansing station was set up at Aguino on the Ria de Arosa with a capacity of 200 tonnes a day (Anon., 1970d). By siting the station near an unpolluted source of sea water there is no need for artificial purification by ultraviolet light. The cleansing tanks were constructed from cement faced brickwork, 22 m long by 0·75 m wide, with a sloping floor from a depth of 175 cm falling to 375 cm, Fig. 135. The mussels are placed on nylon netting supported by short lengths of

wood just below the water surface, the depth of water below them being sufficient to allow the mussels' faecal material and detritus to be carried away by the water currents.

The cleansing time for mussels using this system is between 12 and 24 hours, a considerable saving in time compared with other methods.

In Chapter 4 it was mentioned that the water quality at the site for a future farm should be studied to ensure satisfactory water supplies. This assumes a knowledge of what the criteria are for selection. One recent unfortunate example of the choice of location for a pilot oyster hatchery in Britain has come to light. This unit was established under controlled commercial conditions, using methods developed by the Fisheries Experimental Station of the Ministry of Agriculture, Fisheries and Food at Conway (Walne, 1966). After three unsuccessful seasons at the pilot hatchery at Conway to obtain a good settlement of larvae from their breeding stock of oysters, *Ostrea edulis*, the water quality was questioned. An independent geochemical study of the estuarine water used by the hatchery has indicated the existence of lead and zinc

Fig. 134 Details of perforated plastic hose and concrete block weights. (*Credit—P. H. Milne*)

Fig. 135   Mussel cleansing tanks at Ria de Arosa in Spain.   (*Credit—by courtesy of World Fishing*)

(Elderfield *et al.*, 1971). These two heavy metals are no doubt due to the mineralisation and mining activity in the Welsh hills surrounding the estuary.

Subsequent preliminary experiments have been carried out at Conway to test the effect of varying amounts of zinc on larval developments of *O. edulis* (Walne, 1970). These tests showed that at concentrations of less than 100 μg Zn/l the effect was minimal, while at 300 μg Zn/l the larval growth rate was considerably reduced and at 500 μg Zn/l the larvae either died or failed to metamorphose.

It would therefore appear that the reason for the poor larvae settlements at Conway has been the zinc concentrations in the hatchery water being toxic to the oyster larvae.

In an expanding industry such as marine farming where there are escalating costs of imported fry for fisheries and seed for shellfish production, it is imperative to consider all aspects of the breeding environment to ensure clean unpolluted water supplies.

5
THE FUTURE

Chapter 14
Recent Advances in Materials and Techniques

Previous chapters have made passing reference to the introduction and deployment of various materials and new techniques. To assist the prospective marine farmer in the selection of farm facilities, the advantages of these methods and the introduction of new materials and techniques are reconsidered and amplified.

Shore Facilities

Previously shore tanks and hatchery facilities for sea farming have been generally constructed in concrete. The Japanese (Williamson, 1971) now regret this choice since many of their concrete tanks, designed for shrimp farming, are now obsolete due to the rapid development of sea farming technology. Hence it is recommended that future marine farmers curtail concrete construction in favour of light construction or Mecanno-style frameworks, which may be easily dismantled and adapted for future projects. In order to simplify such changes and alterations to layouts, plastic and fibreglass tanks have been employed for some time in the United States (Marvin, 1964).

Recently the author completed a survey of sectional tanks for sea farming (Milne, 1972a). This survey was initiated as a result of numerous enquiries regarding flexible shore construction techniques.

The normal steel sections employed for fresh water storage were not considered suitable for sea water and attention was focused on fibreglass or similar materials. Fibreglass bolted panels are available in 1 m square sections and for small tanks are quite suitable, Figs. 49, 136, but are more awkward with large tanks where external bulkheads are required to withstand the hydrostatic water pressure on a long series of panels. In addition, these panels are most expensive, and are very time-consuming in their erection and jointing.

A wood and fibreglass composite material called Timbaglass has several advantages over the previous small panels as large sections up to 3 m in length may be constructed. The water face is coated with fibreglass, and the external timber face weather-proofed. Jointing is by means of an impact adhesive. This material offers a considerable reduction in cost over the fibreglass panels, but the life of the timber is still an unknown quantity, probably 10–15 years, depending on the number of times it is reused.

Fig. 136 Glass-fibre reinforced plastic sectional tanks in 1 m square panels.

(Credit—by courtesy of Nicholson Plastics Ltd)

Raceways, popular for farming rainbow trout, are eminently suitable for either fibreglass or Timbaglass construction, and it is suggested that a "U"-section would be an ideal choice. The projected costs of this "U"-section from manufacturers compares very favourably with sectional panels.

A new sectional tank to appear on the British market in 1971 was introduced by Vortex Ltd and the units are made of expanded polystyrene (Milne, 1972a). Sixteen sections lock together like cheese wedges to form an 8 m diameter circular Swedish tank, Fig. 137, and are held together by three retaining straps round the outside of the tank; no nuts and bolts are required. The tank sections are sealed with a closed cell polyethylene cord—18 mm nominal diameter. These polystyrene units provide an ideal sectional circular tank for sea farming as the polystyrene is both non-toxic and easily cleaned, Fig. 138.

The centre section of each tank contains a rotating disc with three outlet screens of different mesh sizes, 6 mm, 12 mm, and 18 mm, to accommodate various sizes of fish. There is also a clear section to facilitate emptying and cleaning the tank. The tanks can be made predator proof by employing a 25 mm mesh nylon net suspended from a centre pole and tied round the periphery of the tank, as shown in Fig. 138.

Where low lying land may be easily excavated for pond culture, by earth moving equipment, the use of polythene or rubber liners has definite advantages over construction in concrete. These

liners had been used previously for fresh water culture, Figs. 43–5, 70, and have now been successfully used in sea farming in Florida, Figs. 32–4, and in Britain, Fig. 36.

Sublittoral Techniques

The extension of sea farming into the sublittoral zone of coastal waters has necessitated the design of light piled structures which will not substantially modify the previous hydrographic conditions. The choice of a simple galvanised scaffolding framework, Figs. 84, 103 (Milne, 1969b, 1970b, e), and the use of galvanised weldmesh, as a result of marine fouling experience in British and Japanese waters, produces an effective fish retention seal or enclosure. The third and fourth fish enclosures to be constructed using the galvanised scaffolding technique are due for completion in 1972.

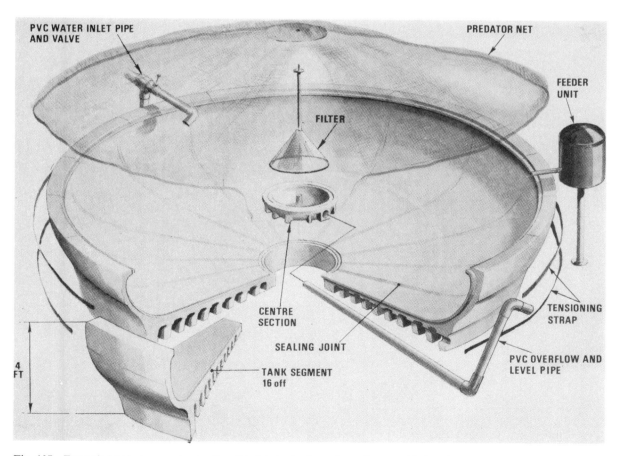

Fig. 137 Expanded polystyrene circular Swedish fish tank. It can be easily assembled and re-assembled from 16 manageable sections using a polyethylene sealing strip. (*Credit—by courtesy of Vortex (Fishery Equipment) Ltd*)

167

Flotation Units

With a widespread and increasing interest in floating cages and rafts for sea farming, many different floating techniques have been developed. Polythene buoys and drums are now used universally as flotation units, Figs. 24, 106, 110, 116, 117, and where the sinking of a platform would entail complete loss of stock, it is essential that they are filled with expanded polystyrene to ensure adequate flotation even when damaged. Styrofoam cylinders are ideal for this application, and have been used in many areas.

The buoys or drums have still to be connected together to form a platform from which to work, and this has led to the development of a one-piece construction of a floating collar, and these are used in Britain and Japan for fish farming and in Canada for oyster farming, Figs. 16 and 107.

However, the fouling of these plastic drums is often considerable, and to simplify the cleaning of the drums, the Japanese ensheath them in a polythene bag which is replaced every year.

Plastic and polystyrene materials are really only suitable for small structures, yet Hunter and Farr (1970) have shown the possibility of assembling small sections into a large floating enclosure extending to 50 m × 12 m, Figs. 119–121.

Lightweight ferro-cement containers may now provide a flotation unit with anti-fouling properties, especially where marine fouling is very severe and *Teredo* is a major problem. The Japanese have already started to use hollow concrete drums for flotation, and it is reported that commercial mussel farming units in Australia contemplate the use of ferro-cement pontoons.

Automatic Feeding Units

Previously it was thought that trash fish was the most suitable food-stuff as it could be converted into edible protein. But the evidence from large scale commercial farms which have purely used trash feeding is that the waste products from this method of feeding build up on the seabed with an increase in sulphides and a decrease in dissolved

Fig. 138 Expanded polystyrene circular fish tank complete with nylon mesh predator net. (*Credit—P. H. Milne*)

oxygen. There has therefore been considerable research into the adaptation and development of pellets similar to those used in fresh water fish farming (Campbell, 1969). On large farms the use of pellets is an advantage since automatic feeding systems can be adopted with a considerable saving in manpower over hand feeding. One such compressed air automatic feeder was put on the British market by Vortex Ltd in 1971 (Milne, 1972b). The feeders can be supplied with single or twin hoppers, Fig. 139. The length of feeding time and feeding interval can be electronically adjusted as required. A bypass switch is also incorporated in the electrics to enable the operator to preset the machine and check the required food quantity, which is blown over the water surface by compressed air. A simple electrically-operated feeder has also been designed recently for use in salmon rearing (Minaur, 1971).

Fig. 139 Compressed air automatic feeder for dry pellets. The feeders can be supplied with single or twin hoppers.

(*Credit—P. H. Milne*)

New Cultch Materials

The collection of natural spat in the sea for shellfish farming has followed traditional lines for many years. Recently the development of lightweight polymeric materials, has simplified this procedure. As mentioned, already, Netlon can now be used for oyster and mussel settlement, for the containment of the mussels on the bouchots, as a fencing material and for stretcher racks and trays (Nortene, 1969). Recent work by Marshall (1970) in a North Carolina tidal creek has studied the use of various cultch materials for oysters. The results of the tests showed that Inertol-coated plastic cultch plates closely spaced together were very satisfactory for the collection of seed oysters. Asbestos cultch plates, however, proved superior for growing the oysters to commercial size.

Adaptation of Heated Effluents from Power Stations

The potential of heated effluents from thermal power plants in extending the growing season of fish and shellfish for commercial sea farming is one of the most exciting prospects for the future (Nash, 1968, 1970; Gaucher, 1970; Beauchamp *et al.*, 1971).

Some commercial and scientific progress has already been made on the evaluation of its potential; oysters are farmed in a heated lagoon beside a power station in Long Island Sound, and research into pompano and shrimp culture is being carried out in shore ponds at a power station in Florida, both in the United States. In Britain research into the cultivation of plaice and sole has been carried out and an experimental pilot farm for prawns has been constructed using nuclear power station heated effluents. The choice of nuclear power stations in preference to coal or oil is that the former have a more reliable and less fluctuating supply of heated water. These projects have already been discussed in detail in previous chapters.

Recirculated Water Systems

With increasing pollution in our estuaries and coastal waters, especially near centres of population, the most suitable sites for sea farming will in the future lie in underdeveloped areas. However, where a regulated salinity supply from fresh to salt water is required for the spawning of some species, it may not be possible to locate areas with an adequate fresh water supply for the hatchery. In these circumstances the use of a recirculated system may possess considerable advantages and increase the potential exploitation of coastal areas.

Considerable research, mostly in the United States, has been carried out into the possibility of using a recirculated sea water system (Parisot, 1967; Wise, 1970). In general most of these systems use part make-up water to avoid a build-up of metabolic products (Brockway, 1950). The advantages of having complete control in a closed system are discussed by Spotte (1970) with reference to the metabolic effects and physiology of the cultivated species.

One factor in the use of heated effluents from power stations, which is seldom appreciated is the poor quality of the water in some areas due to variable salinity and sediment in suspension. The combination of a recirculated system using heated water therefore merits consideration, and one such plant was constructed by the Unilever Research Laboratory near Aberdeen in Scotland for the rearing of 40,000 Dover Sole (*Solea* sp.), to beyond metamorphosis (Phillips, 1970). This system was not completely closed as only 95% of the water was recycled and 5% was supplemental sea water.

Potential Sublittoral Enclosure Techniques

Many references to sea farming in the future refer to the use of electrical, pneumatic and ultrasonic barriers for fish retention. It appears that the numerous experiments which have been undertaken on the effect of such barriers on fish, have not yet produced a completely reliable method that the sea farmer could safely employ for the retention of his stock and the exclusion of predators.

Electrical Barriers Electric fish screens have been used with limited success in Britain, Japan, the United States and the Continent of Europe. Most of these have, however, been designed for use in fresh water, Fig. 140 (Lethlean, 1953; Simpson, 1966; FAO, 1966). The use of electricity in marine waters has mostly been confined to electro-fishing and research has been carried out in Germany by Kreutzer (Houston, 1949) and Dethloff (1964); in America by Harris (1953) and Elliot (1969); and in Japan by Kuroki (1964). In electro-fishing, electrical currents are used for a short time to produce

Fig. 140 Electrical fish screen at outlet from Inverawe Power Station in Scotland, with two parallel rows of vertical electrodes 3 m apart. The downstream row consists of 30 cm diameter aluminium tubes at 2 m centres and the upstream row of 5 cm diameter aluminium tubes at 30 cm centres. (*Credit—P. H. Milne*)

electro-taxis in the fish, so that it orients itself in the direction of the electrical field and swims into a waiting fishing net. The power required to operate this equipment at sea is exceptional, and its use for the creation of sea farm barriers has obvious drawbacks. Nevertheless electro-fishing has definite advantages for removing fish from holding enclosures. Different sizes and species of marine fish require various voltages and currents, and work has been carried out on this aspect by Bary (1956) and Halsband (1959).

Pneumatic Barriers Although air bubble curtains or pneumatic barriers were patented in 1907 for wave protection, it was not until the nineteen fifties that they were considered for use in the fishing industry. Since then they have been suggested on numerous occasions for use in sea farming (Bardach, 1968).

The first apparent use of pneumatic barriers by the fishing industry was reported by Radionov (1958) in Russia, where pneumatic barriers were used to prevent fish from being thrown out of the nets and to preserve the nets from damage in bad weather when being hauled inboard.

The Japanese Fisheries were probably the first to study the effect of air bubble screens on fish. Kobayashi *et al.* (1959) made preliminary observations on the behaviour of a fish school in relation to an air bubble screen. Enami (1960a, b) then extended these experiments to form an air bubble net in order to retain fish.

The success of model experiments with pneumatic barriers prompted the Maine Fishing Industry in Canada to investigate the usefulness of air bubble screens for fish catching. These pneumatic barriers were supplied air from compressors mounted on trawlers and a perforated pipeline was towed in front of the fish to divert them into traps and stake nets set along the shore.

Twelve commercial units were constructed in 1960 (Smith, 1964) and were not only applied to the above situation, but were also used to herd shoals of herring and menhaden from deep water into shallow water where conventional fishing methods could be used. Smith concluded that air bubble curtains were ineffective in turbulent conditions, or where currents exceeded 1·5 m/s.

Since the early model tests carried out in Japan, on the reactions of fish to air bubble curtains, little more work was done until 1965 when simultaneous investigations were conducted in Norway and Scotland. In Norway the studies showed that an air bubble curtain would stop fish just as effectively

as a fishing net (SINTEF, 1965). In one experiment it is reported that 50 coalfish were herded into the corner of a 3 m deep water tank by an air bubble curtain, and not even a diver could scare the fish through the barrier. It is now hoped to make this technique a practical proposition for commercial fisheries. The Norwegian fiords are believed to be especially suited for this new method of herding, whereby a pneumatic barrier placed at the entrance, can be switched on to prevent the escape of one of the large seasonal shoals of fish once they have moved into the fiord.

A series of experiments were also conducted by Blaxter and Parrish (1966) at the Marine Laboratory, Aberdeen, Scotland. These were fundamentally herding experiments, and it was found that air bubble curtains were only effective if moved slowly, so that the bubbles did not stream back obliquely. This corroborates the findings of Smith for conditions in fast flowing currents.

Hence, with certain limitations, it would appear possible to retain fish with an air bubble curtain, but the problem of predators requires further study, as very little research has been carried out on larger aquatic animals. Kenney (1968) has reported that in the protection of beaches from sharks, pneumatic barriers were tried without success. Seals can also swim through the air bubble curtains. It appears, therefore, that air bubble curtains could only be recommended if suitable methods were also used for the exclusion of predators.

In assessing the potentialities of pneumatic barriers for fish retention, air bubble curtains appear to form suitable barriers in quiet flowing waters under 1·5 m/s, Fig. 141(a). The wave reducing effect of the air bubble curtain is beneficial as it will reduce water turbulence and wave height within the enclosed area (Milne, 1970a), Fig. 141(b), and also increase the dissolved oxygen content. Several disadvantages do, however, exist since air bubble curtains have successfully been applied to arrest the salt water wedge effect in estuaries and at canal lock gates (Larsen, 1960; Abraham *et al.*, 1965; and Zuidhof, 1966) and with regular use air bubble curtains could modify the salinity within the enclosures, Fig. 141(c). Air bubble curtains can also be used for oil spill containment, Figs. 133, 134, 141(d).

The water temperature can also be affected by the upwelling of bottom water. In winter the upwelling of warm bottom water has been used to melt surface ice (Ince, 1964; Anon., 1966c, 1970a) but with the reduction of bottom water temperatures due to recirculation. Similarly the use of air

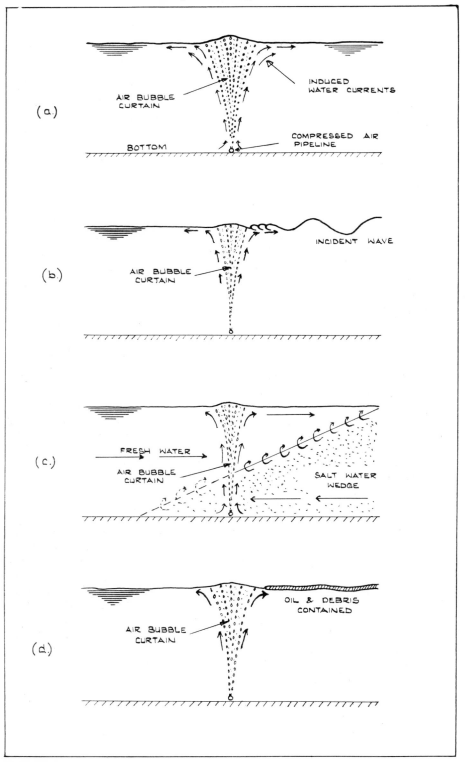

Fig. 141 Air bubble curtains or pneumatic barriers:
    (a) Used to induce water currents.
    (c) Used to prevent the incursion of salt water.
    (b) Aimed at reducing wave height.
    (d) Intended to contain oil and debris.
(Credit—from Milne, 1970a)

bubble curtains in summer will bring colder water from the depths and lower the surface sea water temperature.

Hence, with so many adverse environmental factors involved in the use of pneumatic barriers, they cannot be generally recommended (Milne, 1970a). Yet in exceptional circumstances where salinities and temperatures are uniform from top to bottom, air bubble curtains could be employed, but each case would have to be carefully investigated to ensure that it was not detrimental to the hydrographic environment.

Sonic and Ultrasonic Techniques Many interesting suggestions have also been made that sonic and ultrasonic devices could be used for the herding and farming of fish (Bardach, 1968).

The effects of the mechanical noises of trawl gear and the attendant fishery vessel have received considerable attention in an attempt to increase fish catching. Work by Chapman (1964), with reference to trawls, and Blaxter and Parrish (1966), on the reaction of herring in a tank, has helped to indicate the frequencies at which the fish are stimulated to give a response to mechanical noises under water. A paper by Freytag *et al.* (1969) has focused attention on the problems of mechanical noise on fishing. Noises which disturb fish, causing them to swim away from the vessel and trawl, are thus of interest for fish retention methods to create sonic screens to repel the fish, for example at the sea entrance to an enclosure.

Some Russian experiments (Novosti, 1967) specifically studied the sounds of fish eating, and then by producing these sounds with ultrasonic emitters, shoals of fish were brought swimming into waiting nets. Similarly, reproduced sounds of frightened fish scattered the shoals. One problem to arise was that predators, such as sharks, on hearing the sound of either fish eating or frightened fish, would come to investigate.

This technique of Pavlovian reflexes for the feeding and harvesting of farmed fish is at present under study in Japan. These experiments were commenced by Dr Uchihashi (Bardach, 1968), but he has now retired and his work is being continued

by Dr Fujiya (Bardach, Personal communication). The technique adopted is to sound a particular signal to attract the fish to the automatic feeding devices. The sound is emitted just before feeding and again during feeding, and the food is dispensed by means of an automatic rotating pellet-broadcasting device, which spreads the pellets over a wider area and more evenly than by manual distribution. It has been found that once the fish are trained to respond to the sound signals the net barrier at the entrance to the enclosure can be removed, as the fish will now respond to their "food bell" even when outside the enclosure. When the fish have reached marketable size they can be caught by surrounding them with a seine net at feeding time. These experiments started with trout, and they have now been extended to red sea bream. It is proposed to farm them in areas without the need for net barriers (Fujiya, Personal communication).

Sonic barriers, to be applicable generally for fish farming will require not only a knowledge of the sounds necessary to attract and repel the farmed fish, but also the essential knowledge of the sounds to repel predators.

Artificial Upwelling of Deep Sea Water

In areas of the world's oceans where upwelling of bottom water occurs, there is abundant fish life to support a thriving fishing industry, as for example the Peru Current. This upwelling of nutrient-rich water to the surface leads to an enormous increase in fish growth (Hela and Laevastu, 1970).

This aspect of sea farming is at present being investigated at St Croix in the Virgin Islands by the Lamont-Doherty Geological Observatory of Columbia University (Pinchot, 1970). At St Croix a 9 mm diameter plastic pipe extends nearly 1·5 km into the Caribbean, enabling water to be pumped from a depth of 950 m at a temperature of 5°C. The research so far has shown that selected plant life from the sea water off St Croix grows 27 times faster in ponds using the deep water than surface water. Further experiments are planned to explore the possibilities of upwelling for sea farming.

Chapter 15

Conclusions

The prospective fish and shellfish farmer of the nineteen seventies, has a wide choice of species to cultivate and an ever increasing variety of materials and methods at his disposal. This book has attempted to cover as many as possible of the latest developments in design and construction techniques used for sea farming in coastal waters. From a survey of the literature cited and illustrations used it will be seen that the majority of the papers have been written since 1965 due to the intensification of marine research by government, university and industrial laboratories throughout the world.

The expansion of fish and shellfish farming in coastal waters will only be limited in the future by the quality of the sea water available. As marine pollution has seriously affected many estuaries near industrial zones and centres of population, governments are aware of the need to control marine discharges to prevent the loss of our estuaries and coastal zones as breeding and spawning grounds for young fish and shellfish. The prospective sea farmer should therefore choose his area for farming carefully, preferably away from industrial belts where there is danger of marine pollution to ensure a satisfactory sea water supply.

In assessing the potential of a site the prospective sea farmer now has extensive advice as to the enquiries necessary at the preliminary planning stage. After ascertaining the requirements of the local legal conditions, hydrographical and hydrological surveys are essential to ensure satisfactory sea water salinities. The exposure of the coastal waters to waves and tidal currents must also be determined to ensure adequate protection during storms.

In the selection of stock for farming it is wise to choose an indigenous species since the marine biologists of the country concerned should have some knowledge of the biology and ecology of the species. The hydrography of the area chosen must also provide a suitable environment for the growth of that species. If starting farming a species for the first time in a new country, having read and observed the spectacular success with the same species in a neighbouring country, the prospective sea farmer must appreciate that seldom can farming methods be transferred without modification. It is inevitable that the climatic and hydrographic conditions will be different, and as a result the previous techniques will require to be adapted to suit the new location.

In the selection of species for sea farming it should be appreciated that in mollusc farming where spat is planted on an intertidal beach or in a shallow sublittoral zone the farmer is using the environment to his advantage. Here the molluscs are filter feeders and their food supply is provided by the phytoplankton in the sea water tidal currents. The mollusc farmer, after laying the spat, therefore has little more to do than monitor the beds for predators, and redistribute the molluscs to promote better growth prior to harvesting.

If, however, the sea farmer should choose a species of fish for pond farming, or a floating method of mollusc farming, he must appreciate that these more complicated types of sea farming, although yielding greater harvests, also require more care and skill for their successful growth. It is therefore essential that the prospective sea farmer read all the literature he can to ensure that he has left no gaps where he may have to resort to trial and error methods since these are rarely successful.

One of the major obstacles that the sea farmer will encounter is that of finance. The prospective marine farmer must appreciate that at the outset he needs to have the necessary financial backing with long term loans to support his venture since it is often some time before the first stocks are harvested. For example, one salmon farming concern was established in 1965 and after first developing hatchery techniques for the supply of smolts, and second the construction of sea enclosures which were stocked for the first time in 1969. It was not until 1971, however, that the first salmon were sold to the market, showing the possible length of time required before there is any return for the farm.

In recent years considerable attention has been paid to the potential use of heated effluents from coastal power stations for the artificial culture and improvement in growth rates of many warm water species. One of the most exciting developments in this sector is the possibility of commercial lobster farming since the use of heated water enables marketable lobsters to be grown inside three years compared to five years in the sea.

The development in the last decade of surface floating units for both rafts and cages in the sea has been one of the major aspects in the expansion of sea farming. This technique has increased the potential available space for sea farming and has brought it into the realm of small crofting communities working within a co-operative organisation as in Japan.

The potential for the economic cultivation of fish and shellfish in coastal waters is now so great that we stand on the verge of a world wide sea farming industry. This provides a genuine challenge to those who are prepared to work hard to understand the biological, environmental, engineering and managerial factors involved in running a successful and financially rewarding enterprise.

Appendices

These five Appendices are taken from a previous paper by the author, entitled, "Fish Farming: A Guide to the Design and Construction of Net Enclosures" (*Marine Research Scotland*, 1970 No. 1), and they are reproduced by kind permission of the Controller of Her Majesty's Stationery Office.

Appendix 1

Wind Forces on Coastal Structures

For the design of permanent structures to form enclosures in coastal waters, the dynamic load, due to wind on the structure, must be assessed. This loading is subjected to a "probability" of occurrence, and has to be evaluated from the extreme meteorological conditions pertaining to the area considered.

The author's research work on sea farming enclosures has centred on the west coast of Scotland and so the following meteorological conditions for sea lochs, coastal areas and outer isles are initially applicable to this area. However, the figures can be used generally by obtaining from the local meteorological office details of the extreme wind speeds pertaining to the area of interest and entering the figures with these local wind speeds.

The British Meteorological Office report on extreme wind speeds (Shellard, 1965) presents two figures showing the highest mean hourly wind speeds and general distribution of gust speeds at 10 m above the ground, likely to be exceeded on average only once in 50 years. This long term value has been chosen with due consideration to the long term nature of sea farming.

For the design of marine structures the gust speed is the most critical design loading and is used to calculate the wind forces, F (Boven, 1968):

$$F = 0.0965AV^2 \qquad \text{(Eqn 1)}$$

where

F = wind force in kg
A = area in m²
V = velocity in m/s

Equation 1 gives the wind force on the solid parts of the structure, but for the portions of netting round the enclosures a lower figure can be expected since the nets do not present such an obstacle to the wind.

In Japan, Tamura and Yamada (1963) conducted experiments in a wind tunnel on fish netting and for the calculation of wind force, F, they present the following:

$$F = 0.0186AV^2 \qquad \text{(Eqn 2)}$$
(same notation as for Eqn 1)

However, equation 2 makes no allowance for different types of netting and mesh sizes. A comparison with British wind tunnel technique (Pankhurst and Holder, 1952) for the resistance of gauzes gives:

$$F = \tfrac{1}{2} K \rho V^2 \times 10^{-2}$$
$$\text{(Eqn 3)}$$

where

F = wind force in kg
K = resistance coefficient $= \dfrac{1-\beta}{2}$

Area	Mean hourly wind speeds mile/h	m/s	Gust speeds mile/h	m/s
Coastal and sea lochs	70–80	31–36	110–120	49–54
Offshore and outer isles	80–90	36–40	120–130	54–58

176

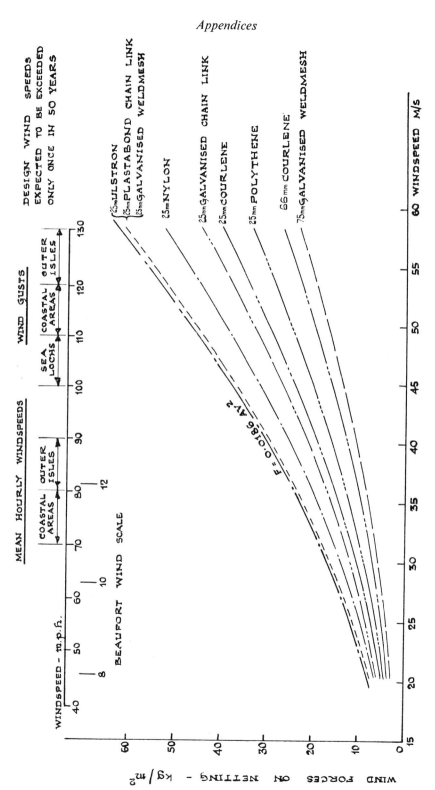

Fig. 142   Wind forces on mesh netting for various wind-speeds. (Wind speeds after Shellard, 1965; F = 0·0186 AV², after Tamura and Yamada, 1963.)   (*Credit—from Milne, 1970e*)

β = blockage coefficient $= (1-d/l)^2$
d = mesh diameter in m
l = nominal mesh size in m
ρ = density of air $= 1.227 \text{ kg/m}^3$
V = velocity of wind in m/s

The wind force on several types of mesh fabric, synthetic and metallic for various wind speeds are presented in Fig. 142 and are compared with the wind force given by Tamura and Yamada in equation 2. The Japanese figures are practically identical with the forces on 25 mm Ulstron, Plastabond chain-link and galvanised square mesh but are excessive when compared with other types of netting and mesh sizes. Fig. 142 therefore, gives the design on various types of clean netting for a given wind speed depending on the locality. If the netting is liable to foul with drift weed in the intertidal zone, consideration should be given to the forces on the percentage of the net fouled and these should be calculated from equation 1.

If two nets are inserted side by side, say a 25 mm fish retention net with a 75 mm predator and trash net, as described for Faery Isles in Scotland, the net structure must be capable of withstanding the combined wind forces of the two nets.

Appendix 2
Wave Forces on Coastal Structures

Coastal and offshore structures are normally designed to withstand a "highest" single wave pertaining to the average wave heights occurring with a small probability. The design wave height has to be calculated from:

1. the characteristics of the wind field;
2. the direction and speed of the wind field;
3. the fetch length; and
4. the water depth variations along the fetch.

This means that any site for a possible enclosure has to be investigated separately depending on its geographical location and topographical features. From the previous section on winds, we know the highest mean hourly wind speed which can be expected in the area, so for a given fetch length it is possible to predict the height and period of expected waves from forecast charts such as those presented for easy use in "Shore Protection Planning and Design" (U.S. Army, 1966). As indicated above the wave height is dependent on the fetch length and water depth variations along the fetch. In Fig. 143 curves are drawn giving the wave heights for fetch lengths under two conditions, firstly in deep water and secondly in shallow water (6–15 m). Three curves are drawn for each condition:

i.	force 8 gale ..	..	..	..	20 m/s
ii.	force 10 storm ..	..	..	..	28 m/s
iii.	force 12 hurricane ..	..	..	36 m/s

The wave height will depend on the location of the enclosure and the depth of water offshore, but for exposed sites on the coast the deep water wave height should be taken, and for sheltered inlets the shallow water wave height.

In a narrow fiordic sea loch the width of the fetch may have a limiting effect on the generation of waves. Here the ratio of fetch width to length is very important, which may result in the generation of waves considerably lower than those that would be expected under the same generating conditions in open waters. The fetch effectiveness for various ratios of fetch width to length is given in Fig. 144 from a paper by Saville (1954). Due attention, however, should be given to the funnelling effect of this situation on a loch with surrounding steep hills which may not allow any reduction in wave height to be made.

To calculate the wave forces on netting the orbital velocities of the water particles within the waves, height (H), are required and these vary depending on the depth of water (d), the period (T) and wave length (L). To calculate the orbital velocities, Stokes' theory for waves of finite amplitude, as presented and discussed by Wiegel (1964), is applicable down to $d/L = 0.125$, which covers depths within the area of interest. The maximum horizontal orbital velocities for wave heights up to 2 m in 6 m and 15 m of water are presented on Fig. 145 for periods ranging from 2.5 to 8 seconds. The maximum vertical orbital velocities are found by a similar calculation from Stokes' theory and are approximately 82.5% of the maximum horizontal orbital velocity. It will be seen that the limiting condition for wave steepness (Wiegel, 1964)

$$H = 0.45 \ T^2 \qquad \text{(Eqn 4)}$$

causes waves of small periods to break before reaching any height; thus the maximum horizontal orbital velocity for any wave condition for sea farming applications is unlikely to exceed 2 m/s.

Some research work carried out on wave forces on fish netting has been reported by Tamura and Yamada (1963) where:

$$F_h = 2.15 \ u_{max} \qquad \text{(Eqn 5a)}$$
$$F_v = 1.80 \ v_{max} \qquad \text{(Eqn 5b)}$$

where F_h, F_v are horizontal and vertical component forces of wave pressure respectively, and u_{max}, v_{max} are maximum component velocity of horizontal and vertical movement of water molecules on the surface respectively.

However, the above equations 5a and 5b make no allowances for mesh size, so a more reliable estimate

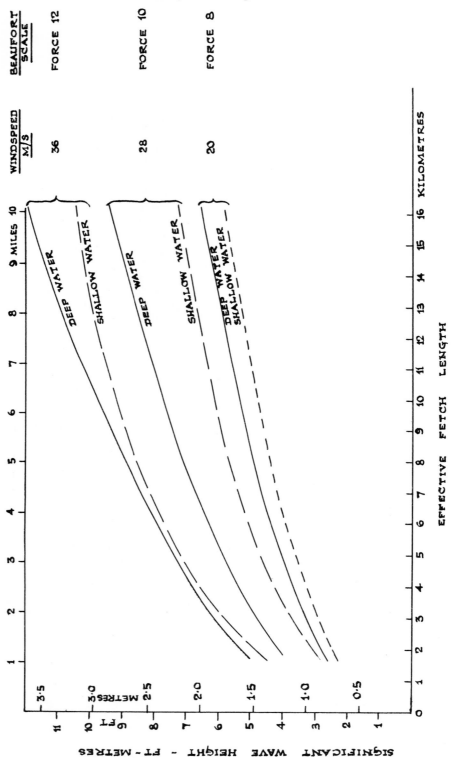

Fig. 143 Significant wave heights for various wind speeds and fetch lengths (wave heights after U.S. Army 1966).

(*Credit—from Milne, 1970e*)

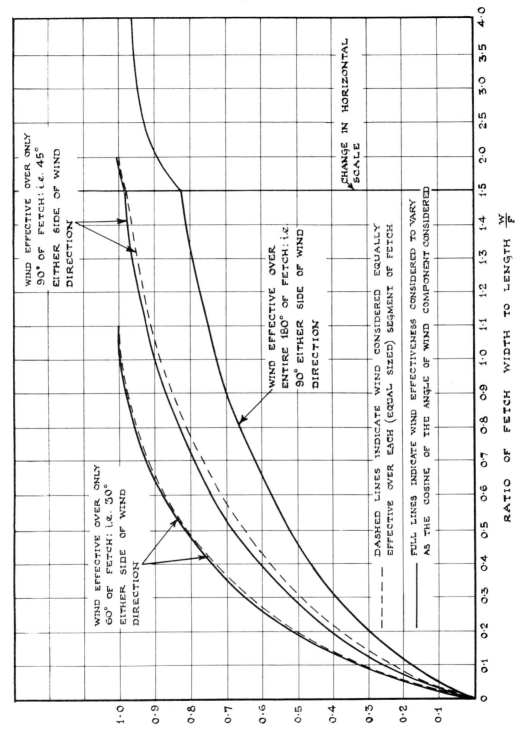

Fig. 144   Relationship of effective fetch to width/length ratio for rectangular fetches (after Saville, 1954).

(*Credit—from Milne, 1970e*)

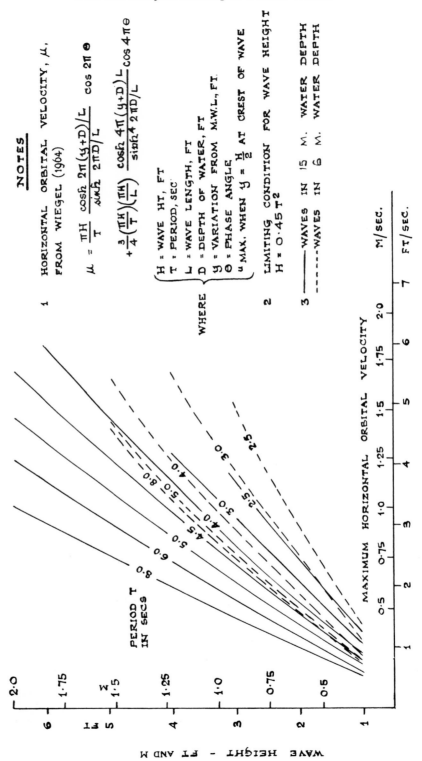

NOTES

1 HORIZONTAL ORBITAL VELOCITY, μ, FROM WIEGEL (1964)

$$\mu = \frac{\pi H}{T} \frac{\cosh 2\pi (y+D)/L}{\sinh 2\pi D/L} \cos 2\pi \theta$$
$$+ \frac{3}{4} \left(\frac{\pi H}{T}\right) \left(\frac{\pi H}{L}\right) \frac{\cosh 4\pi (y+D)/L}{\sinh^4 2\pi D/L} \cos 4\pi \theta$$

WHERE
H = WAVE HT, FT
T = PERIOD, SEC
L = WAVE LENGTH, FT
D = DEPTH OF WATER, FT
y = VARIATION FROM M.W.L., FT
θ = PHASE ANGLE
μ MAX. WHEN $y = \frac{H}{2}$ AT CREST OF WAVE

2 LIMITING CONDITION FOR WAVE HEIGHT $H = 0.45\ T^2$

3 —— WAVES IN 15 M. WATER DEPTH
---- WAVES IN 6 M. WATER DEPTH

Fig. 145 Horizontal orbital velocities in 6 m and 15 m of water for various wave heights and periods.

(*Credit—from Milne, 1970e*)

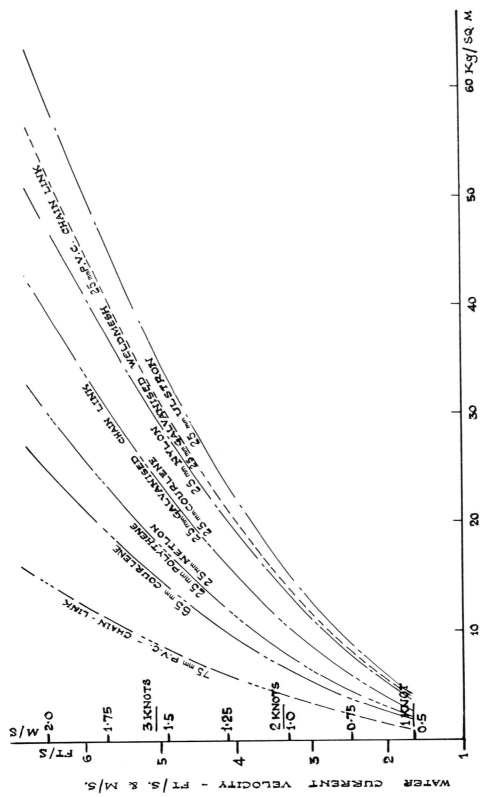

Fig. 146   Current forces in kg/sq. m on clean fish netting panels.   (*Credit—from Milne, 1970e*)

would be to calculate the wave pressure from current drag, equation 9, presented in Appendix 3 for the calculation of forces due to tidal currents, where the mesh size and twine dimensions are taken into consideration. The forces due to water currents passing through mesh screens are presented for several netting fabrics in Fig. 146 where it will be seen that with higher orbital velocities the forces on the netting increase appreciably. The forces given are those for clean netting panels, which in practice would seldom occur. In general, the area exposed to wave action will not be fouled like the under-water section which is permanently immersed, but it will collect some drift weed, and an estimate of the area blocked by drift weed is required. From the author's enclosure investigations at Loch Sween a realistic figure would be to consider 20% of the surface to be fouled, i.e., blocked. The wave force can then be arrived at by considering 20% of the surface solid and the other 80% clean mesh. For a given orbital velocity the wave force presented in Fig. 146 for clean mesh should be doubled for 20% fouling to allow for surface drift weed and fouling.

Appendix 3
Tidal Current Forces on Mesh Netting

If the enclosure is sited in a coastal environment it will be subjected to a daily tidal cycle which will set up associated tidal currents. The design and orientation of the enclosure should be settled only after a hydrographical survey has been carried out to determine the strength and direction of these currents.

Research into the resistance of fishing nets to currents was started by Terada *et al.* as far back as 1914. Since then, numerous papers have been presented on different aspects examining various nets, flows and approach angles. More recent papers by Kawakami have given equations for general use. His first paper (Kawakami, 1959) presented the relationship:

$$R = kSU^2 \qquad \text{(Eqn 6)}$$

where

R = resistance of net
S = area of webbing
U = velocity of current

and

k = coefficient of resistance depending upon the construction of the net and angle of attack.

In a further paper Kawakami (1964) examined the resistance due to flow at right angles to a net and presented a more useful equation:

$$R = \frac{C_d \rho v^2 S}{2}. \qquad \text{(Eqn 7)}$$

where

R = resistance of net
C_d = coefficient of drag of mesh
ρ = density of water
v = velocity of current
S = projected area of net = $2ad$

and

a = nominal mesh size
d = diameter of twine.

Equation 7 has been investigated in a comprehensive British research report by Saunders-Roe in which a very good comparison with theory is achieved. For final derivation of C_d, the twine coefficient of drag for "standard" conditions is given for:

a knotted net:
$$C_d = 1 + 3.77\,(d/a) + 9.37\,(d/a)^2 \qquad \text{(Eqn 8a)}$$
and
a knotless net:
$$C_d = 1 + 2.73\,(d/a) + 3.12\,(d/a)^2 \qquad \text{(Eqn 8b)}$$

From the values of C_d obtained it is then possible to determine the drag per mesh from:

$$C_d = \frac{\text{drag per mesh}}{\frac{1}{2}\rho v^2 (2ad)} \qquad \text{(Eqn 9)}$$

which is the same as presented by Kawakami in equation 7.

It is now, therefore, possible to calculate the anticipated forces on panels of clean fishing net. The forces in kg/m² are shown in Fig. 146 for clean fish netting, for each of the mesh fabrics selected for the University's netting survey, as detailed in Fig. 147. However, no matter how often the fish nets are cleaned, for design purposes it is necessary to consider the nets to be in a fouled state. As a result of the research into the fouling of fish netting (Milne and Powell, 1972), summarised in Chapter 5, it is necessary to consider the forces on fouled fish netting. Fouled data is detailed in Fig. 148 and Fig. 149 as the forces in kg/m² on panels of fouled fish netting. This research confirms the work by Nomura and Mori (1956) who found that the drag of Japanese fishing nets depended not only on the physical dimensions but also on the texture and material.

Where two nets are inserted side by side, say a 25 mm fish retention net with a 75 mm predator and trash net, the net structure must be capable of withstanding the combined water forces of the two nets.

COMPARISON OF VARIOUS TYPES OF NETTING
(Numbered to Correspond with Marine Fouling Test Panels—Figs. 19 and 20)

Net Ref No.	Net Material	Mesh Type D = Diamond S = Square	Nominal Bar Length "a" in.	Twine Diameter "d" in.	"a/d"	C_d
1	Nylon	D	1	0·09	11·1	1·416
2	Ulstron	D	1	0·10	10·0	1·470
3	Courlene	D	1	0·075	13·3	1·332
4	Polythene	S	1	0·06	16·6	1·260
5	Polythene (Cupra-proofed)	S	1	0·06	16·6	1·260
6	Plastabond Chain-link	D*	1	0·10	10·0	1·304
7	Galvanised Chain-link	D*	1	0·08	12·5	1·238
8	Plastabond Chain-link	D*	3	0·10	30·0	1·094
9	Netlon	S*	2	0·13	15·4	1·192
10, 11	Galvanised Square Mesh	S*	1	0·10	10·0	1·304
—	Courlene	D	2·65	0·138	19·2	1·198

* Knotless Net; 2.65 in. Courlene, not tested but shown for comparison.

Fig. 147   Comparison of various types of netting used in marine fouling studies.   (*Credit—from Milne, 1970e*)

COMPARISON OF VARIOUS TYPES OF FOULED NETTING AFTER TWO MONTHS' IMMERSION IN SEA WATER

Net Ref No.	Net Material	Mesh Type D = Diamond S = Square	Nominal Bar Length "a" in.	Twine Diameter "d" in.		a/d_f	C_{df}
				"d" clean	"d_f" foul		
1	Nylon	D	1	0·09	0·40	2·5	3·985
2	Ulstron	D	1	0·10	0·40	2·5	3·985
3	Courlene	D	1	0·075	0·35	2·9	3·455
4	Polythene	S	1	0·06	0·30	3·3	2·953
5	Polythene (Cupra-proofed)	S	1	0·06	0·20	5	2·129
6	Plastabond Chain-link	D*	1	0·10	0·20	5	1·671
7	Galvanised Chain-link	D*	1	0·08	0·15	6·7	1·48
8	Plastabond Chain-link	D*	3	0·10	0·25	12	1·25
9	Netlon	S*	2	0·13	0·30	6·6	1·48
10, 11	Galvanised Square Mesh	S*	1	0·10	0·13	7·7	1·408
—	Courlene	D	2·65	0·138	0·40	6·6	1·78

* Knotless Net; 2·65 in. Courlene, not tested but shown for comparison.

Fig. 148   Comparison of various types of fouled netting after two months immersion in sea water.

(*Credit—from Milne, 1970e*)

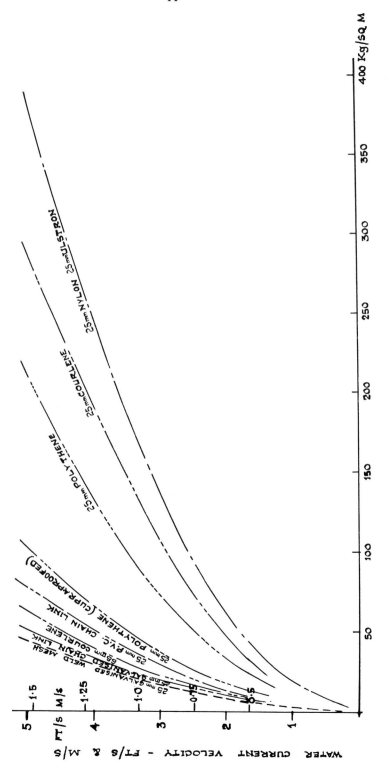

Fig. 149 Current forces in kg/sq. m on fouled fish netting panels. (*Credit—from Milne, 1970e*)

Mesh Net Design Criteria

The main forces on any net structure, apart from superimposed design loadings for maintenance and access, are the variable forces arising from wind, waves and currents already discussed in Appendices 1–3. The mesh net design criteria taken here for design assume a wind gust speed of 120 mile/h (54 m/s) a wave orbital velocity of 6 ft/s (1·83 m/s) and a current of 2 knots (1 m/s). For illustration, the forces on three different fabrics are summarised from Figs. 142, 145 and 149.

Design Forces in kg/m²
25 mm Mesh Fabric Material

		Courlene	Polythene	Gal. Weldmesh
1. wind				
	force	33·7	28	52
2. wave				
	force	70	53	90
3. current				
	force	141	98	20

Here it will be seen that the forces on the fibre nets are least at the air-sea interface and greatest below water. However, the greatest problem with fibre nets is their installation to prevent chafing and abrasion owing to continual flapping in the wind if not stretched taut and tied regularly.

To determine the frequency of supports necessary, consider a 1 m square panel of 25 mm mesh courlene (net break 49·5 kg) strung between two supports subjected to a maximum underwater force of 141 kg/m².

The breaking load when new = 2 × 39 × 49·5 = 3860 kg.

If courlene is allowed three years' life with 50% strength at the end of three years and a reduction of 50% to prevent extension:

$$\text{factor of safety} = \frac{3860}{4 \times 141} = 7$$

which could be reduced further for optimum design. A 2 m sq. panel, factor of safety = 3·5, may be regarded as adequate.

For 25 mm polythene mesh (wet break 45·5 kg), consider a force of 98 kg/m² on a 1 m square panel, and allow the polythene three years' life with 50% strength at the end of three years and a reduction of 50% to prevent extension:

$$\text{factor of safety} = \frac{3550}{4 \times 98} = 9 \cdot 05$$

To give a safety factor of 3·5 the size of panel could be increased to 2·5 m square, owing to its higher resistance to marine fouling.

For the 25 mm galvanised mesh (wet break 205 kg) it is the wave force which produces the highest loading of 90 kg/m². For a 3 m square panel, owing to its greater strength, a 70% strength retention after three years is allowed:

therefore factor of safety

$$= \frac{2 \times 117 \times 205}{9 \times 90} \times 0 \cdot 7 = 41 \cdot 5;$$

thus for 25 mm galvanised mesh the maximum distance between supports can be much greater than for synthetic meshes, but for fixing is recommended at 3 m.

Since the netting is considered to be supported only at two edges, only horizontal supports are necessary for the installation of mesh panels. These horizontal supports may be rigid bars or wire hawsers stretching between the main net structure supports.

Pile Design Calculations

The design of a piled structure to provide a frame-work from which nets may be suspended to form a fish enclosure requires a knowledge of the under-lying sediment. The surface layer should not be taken at face value and cores of the underlying sediment are required to determine their composition and location of bedrock or other stable layer on which to found the structure. In sheltered coastal areas and sea lochs the seabed sediment consists mainly of silt and soft clays near the surface with harder clays in underlying layers down to bedrock. Some sandy seabeds are known but generally in areas exposed to wave action.

The properties of cohesive soils, found in seabed sediments are given in Fig. 150 extracted from Table 12 of Reynolds' Reinforced Concrete Designer's Handbook (1961).

For design purposes soft puddle clay has been chosen as it is representative of the softer seabed materials likely to be found in the more sheltered areas for marine enclosure work:

Soft puddle clay data:
saturated density $\qquad \omega = 1920$ kg/m³
angle of internal friction $\qquad \theta = 3^\circ$
cohesion $\qquad C = 3290$ kg/m³

Example (1)

Consider driving piles in an area with a tidal range of 2 m in 6 m of water at low water, and allow 6 m penetration of the seabed. From the section on waves, Appendix 2, allow 2 m freeboard above high water, giving a length of pile of $2 + 2 + 6 + 6 = 16$ m. Using the design loadings obtained for 25 mm galvanised weldmesh i n a 54 m/s wind gust

PROPERTIES OF COHESIVE SOILS

Type of Soil	Saturated Density ω (kg/m³)	Angle of Internal Friction θ	Cohesion (kg/m²)
Clay:			
very stiff boulder		16°	17,500
hard shaley	1920 to 2240	—	> 14,600
stiff		7°	7,300 to 14,600
firm	1760 to 2080	6°	3,650 to 7,300
moderately firm	1760 to 1920	5°	5,450
soft		4°	1,830 to 3,650
very soft		3°	< 1,830
Puddle Clay:	1600 to 1920		
soft		3°	3,290
very soft		0°	2,200
Sandy Clay:			
stiff	1600		7,300 to 14,600
firm			3,650 to 7,300
Silt:	1920		< 3,650

Fig. 150 Properties of cohesive soils. (*Credit—from Milne, 1970e*)

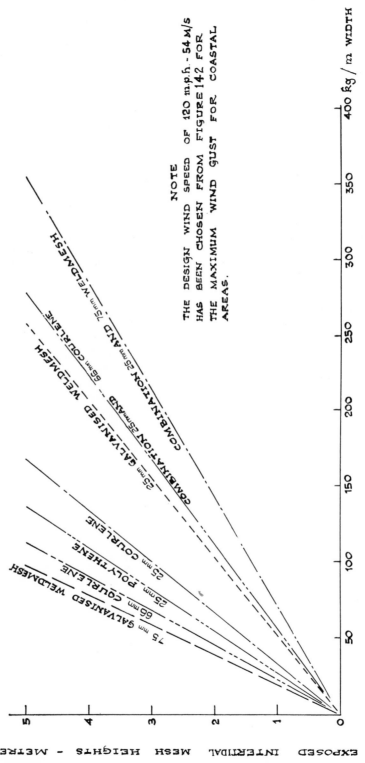

Fig. 151   Design wind forces on mesh netting for various intertidal heights.   (*Credit—from Milne, 1970e*)

with a wave height of 2 m and a current of 1 m/s the total design loadings are in kg/m width of barrier:

	high water	mid-tide	low water
wind force	104	156	208
wave force	180	180	180
current force	160	140	120

Normally at dead high or low water there will be no current forces but these should be considered as if they were present. From a study of the above forces, the pile spacing should be designed to take the maximum loading which occurs at low water. With increasing tidal ranges these wind forces become excessive as shown on Fig. 151. Taking moments about the equilibrium point for active and passive pressures (Reynolds, 1961) and considering only the lower half of the embedded pile to be held solid, a factor of safety of 2·5 gives pile spacings of:

4·7 m for 0·6 m diameter piles
3·5 m for 0·45 m diameter piles

Example (2)

Now consider deeper water with a tidal range of 4 m in 10 m of water at low water, and allow 8 m penetration of the seabed. Allowing 2 m freeboard for wave action gives a 24 m pile. The design loadings for wind, waves and currents are the same as example (1). Also assume the same seabed sediment of soft puddle clay, considering only the lower half of the embedded pile to be held solid. Taking moments about the equilibrium point for active and passive pressures as above with a factor of safety of 2·5 gives a pile spacing of 2·5 m for a 0·6 m pile. This spacing is too close and indicates that another method of enclosure should be attempted. With a firmer sediment, as shown in Fig. 150 for the properties of cohesive soils, the pile spacing would be greater.

The design calculations above refer only to the net section of the enclosure and make no allowance for any catwalks or maintenance platforms or machinery which would impose their own loadings. Also if a boat is to be moored to allow cleaning of the nets this should be taken into consideration.

In general for water depths up to 8 m, piling is a reasonable proposition, but in water deeper than 8 m, as in example (2), where the pile spacing is less than 2·5 m, a moored type of net enclosure is a more reasonable proposition, as discussed in Chapter 8.

Selected References

ABRAHAM, G. and BURGH, P. V.D. (1964). "Pneumatic Reduction of Salt Intrusion through Locks," *Jour. Hydraulics Div., ASCE*, 90(HY1), pp 83–119.

AGERSCHOU, H. A. (1966). "Synthetic Material Filters in Coastal Protection," *Jour. Waterways and Harb. Div., ASCE*, Proc. Paper 2745, Vol. 87, No. WW1.

ALLEN, J. H. and MILNE, P. H. (1967). "Enclosures for Marine Fish Farming," *H. A. Ballinger et al., eds. Proc. Conf. Technol. Sea Sea-bed, Harwell*, 1. London, H.M.S.O., 144–157.

ALLEN, J. H. (1968). "Engineering Aspects of Marine Cultivation," *Proc. Meet. Soc. Underwater Tech.*, London, 5th Nov. 1968, pp 11–17. Also in *World Fishing*, 18(8), pp 42–43.

ANDREU, B. (1968a). "Fishery and Culture of Mussels and Oysters in Spain," *Proc. Symposium on Mollusca*, Part III, India, pp 835–850.

ANDREU, B. (1968b). "The Importance and Possibilities of Mussel Culture," Working Paper 5, *Seminar on Possibilities and Problems of Fisheries Development in South east Asia, German Foundation for Developing Countries*, Berlin (Tegel), September, pp 364–375.

ANON. (1962). "A Polythene Atomic Reservoir," *Muck Shifter and Bulk Handler*, Vol. 20, No. 9, Sept. 1962.

ANON. (1966a). "Nylon for Canal Banks," *Dock and Harbour Auth.*, Vol. 46, No. 545, March 1966.

ANON. (1966b). "Precast Units for Breakwaters (Staempfli System)," *Dock and Harbour Auth.*, Vol. 46, No. 543, Jan. 1966.

ANON. (1966c). "Prevention of Mud, Flotsam and Ice Encroachment by Air Curtain," *The Motor Ship*, 47 (April), p 36.

ANON. (1966d). "Built to Stay Empty," *Chemical Week*, 8th Oct. 1966.

ANON. (1968). "Japanese Oysters in the Philippines," *FAO Fish Cult. Bull.*, 1(1), p 6.

ANON. (1969a). "Oyster Culture," *FAO Fish Cult. Bull.* 2(1), pp 7–8.

ANON. (1969b). "International Nickel, Cupro-nickel Chain-link Fencing for Undersea Enclosures," *Inco-Nickel Bull.* (June), 4.

ANON. (1970a). "Bubble Barrier Keeps Fiord Free of Ice," *International Construction*, 9(1), Jan. 1970.

ANON. (1970b). "Pollution Control Report," Part 1, *Ocean Industry*, Vol. 5, No. 6; Part 2, *Ocean Industry*, Vol. 5, No. 7.

ANON. (1970c). "Commercial Shrimp Farming in Florida," *Under Sea Tech.*, 11(10), pp 21, 26.

ANON. (1970d). "Cleansing 200 tons of Mussels a Day," *World Fishing*, 19(10), pp 30–33.

ANON. (1970e). "N.S.W. to Begin Prawn Farming Experiments," *Aust. Fish.*, 29 (Nov), pp 3–4.

ANON. (1970f). "Oyster Farm Project," *World Fishing*, 19(11), pp 58–59.

ANON. (1971a). "Aquaculture in New Zealand," *FAO Aquacult. Bull.*, 3(2), p 8.

ANON. (1971b). "Turtle Farms in Australia," *World Fishing*, 20(4), p 40.

ANON. (1971c). "Off-bottom Oyster Rearing in Nova Scotia," *World Fishing*, 20(4), pp 40–41.

ANON. (1971d). "Oyster Farming in South Africa," *World Fishing*, 20(5), p 57.

ANON. (1971e). "Aquaculture in New Zealand," *FAO Aquacult. Bull.*, 3(4), p 10.

ANON. (1971f). "Boom Saves Oyster Beds," *World Fishing*, 20(8), p 14.

ANON. (1971g). "Tuna Farming," *World Fishing*, 20(8), p 18.

ANON. (1971h). "French Alarmed at Overfishing," *World Fishing*, 20(8), p 24.

ANON. (1971i). "Australian Mussel Farming Experiment," *World Fishing*, 20(8) p 37.

ANON. (1971j). "Australia Tries Prawn Farming," *World Fishing*, 20(9), pp 99–100.

ANON. (1971k). "Green Turtle Cutlets in Your Local Supermarket this Winter?," *Ocean Sci. News*, 13(40), pp 1–2.

ANON. (1971l). "Oil Containment Barrier Looks Good in Sea Tests," *Ocean Industry*, 6(10), pp 10–11, 21.

ANON. (1971m). "Norway Now Exporter of Oyster Brood," *World Fishing*, 20(11), p 25.

ANON. (1971n). "500,000 Pounds of Farmed Shrimp have been Harvested," *Ocean Sci. News*, 13(50), pp 1–2.

ANON. (1972). "Farming Oysters," *Fish Industry Review*, 2(1), pp 6–8.

ARNE, P. (1938). "Contribution à l'étude de la biologie des muges du Golfe de Gascogne," *Rapp. et Proc. Verb. Comm. Int. Exp. de la Méditerranee*, X1(NS) (Sept.).

ARZANO, R. (1959). "Man-made Fibres," *H. Kristjonsson ed. Modern Fishing Gear of the World*. London, Fishing News (Books) Ltd, 13–18.

ASBECK, B. V. (1964). "Bitumen in Hydraulic Structures," Vol. 1 and 2, Elsevier. Publishing Co. Amsterdam, 1964.

ASPHALT INSTITUTE (1961). "Asphalt in Hydraulic Structures," Manual Series No. 12 (MS-12), Nov. 1961.

ASPHALT INSTITUTE (1962). "Asphalt in Beach Erosion Control Structures," Inf. Series No. 122 (15–122), June 1962.

BARDACH, J. (1968). "Harvest of the Sea," Harper and Row, New York.

BARTON, R. (1970). "How we Pollute our Waters—and How we can Clean Them Up," *World Fishing*, 19(2), pp 21–23.

BARTON, R. (1971). "Pollution–the Unknown Factor in Fisheries," *World Fishing*, 20(12), pp 8–13.

BARY, B. McK. (1956). "The Effect of Electrical Fields on Marine Fishes," *Marine Research Scotland*, No. 1, 1956, H.M.S.O., Edinburgh, 32 pp.

BAYARD, P. (1964). "Fermetures fluviales sur le Rhone," *La Houille Blanche*, No. 4.

BEAUCHAMP, R. S. A., ROSS, F. F., and WHITEHOUSE, J. W. (1971). "The Thermal Enrichment of Aquatic Habitats," *Proc. 5th Int. Water Poll. Res. Conf.*, Paper I–II, 7 pp.

BELL, M. C. (1970). "Fisheries Engineering" in *Marine Aquiculture*," Ed. W. J. McNeil, Oregon State Univ. Press, pp 135–140.

BELLIER, J. (1966). "La precontrainte dans les barrages," *Travaux*, No. 376, May.

BERRY, F. H. and IVERSEN, E. S. (1967). "Pompano: Biology Fisheries and Farming Potential," *Proc. Gulf and Caribb. Fish. Inst.*, 19, pp 116–128.

BLANCHET, C. (1946a). "Formation et destruction par un courant d'eau de massifs en pierres," *La Houille Blanche*, 1, No. 2, March 1946.

BLANCHET, C. (1946b). "Technique de la construction des barrages en pierres lancees dans l'eau courante," *La Houille Blanche*, 1, No. 6. Nov. 1946.

BLAXTER, J. H. S. and PARRISH, B. B. (1966). "The Reaction of Marine Fish to Moving Netting and Other Devices in Tanks," *Marine Research Scotland*, No. 1, 1966, H.M.S.O., Edinburgh, 15 pp.

BØHLE, B. (1970). "Forsøk med dyrking av Blåskjell (*Mytilus edulis* L.) ved overføring av yngel til nettingstromper," *Fiskets Gang*, 56(13–14), pp 267–271.

BØHLE, B. and WIBORG, K. F. (1967). "Forsøk med Dyrking av Blåskjell," *Fiskets Gang*, 53, No. 24, pp 391–395.

BORLEY, J. O. (1912). *Mar. Biol. Ass. Internat. Rept IV*, Cd–6125.

BOVEN, C. J. P. VAN (1968). "Design Considerations for Permanent-type Offshore Structures," *Offshore Europe, 1968*, London, Scient. Surveys (Offshore) Ltd, 51–55.

BOWBEER, A. (1970). "Seed Mollusc Production," *World Fishing*, 19(5), pp 44–45.

BOWBEER, A. (1971). "The Underwater Lobster Farm," *World Fishing*, 20(10), pp 36–37.

BRANDT, A. V. and CARROTHERS, P. J. G. (1964). "Test Methods for Fishing Gear Materials (Twines and Netting)," *Modern Fishing Gear of the World*, Vol. 2, Fishing News (Books) Ltd, pp 9–49.

BRETSCHNEIDER, C. L. (1967). "Storm Surges," in *Adv. in Hydroscience*, Ed. V. T. Chow, Vol. 4, pp 341–418, Academic Press.

BROCKWAY, D. R. (1950). "Metabolic Products and their Effects," *Progr. Fish. Cult.*, 12(3), pp 127–129.

BROOM, J. G. (1969). "Pond Culture of Shrimp on Grand Terre Island, Louisiana, 1962–1968," *Proc. Gulf. Caribb. Fish. Inst.*, 21, pp 137–151.

BRUUN, P. (1953). "Coastal Protection," *Dock and Harbour Authority*, Nov. 1953.

BRUUN, P. (1964). "Revetment for Coastal Protection," *Dock and Harbour Authority*, Vol. 44, No. 520, Feb. 1964.

BRUUN, P. and GERRITSEN, 1964 F. "Dutch and Florida Practices on Revetment Design," *9th Conf. on Coastal Eng.*, Lisbon.

BULL, M. R. (1963). "Growth of Fish Culture in Israel," *Fishing News International*, 2(Oct./Dec.).

BURDON, T. W. (1950). "Annual Report of Fisheries Department, Singapore," 1949, Singapore, Govt. Print. Off.

BURGERS, J. M. (1926). "Grondwaterstrooming in de omgeving van een net van kanalen," *De Ingenieur*, 41 Jaargang, No. 32, August 1926.

BURROWS, R. E. and CHENOWETH, H. H. (1955). "Evaluation of Three Types of Fish Rearing Ponds," *U.S. Dept. Interior, Fish and Wildlife Ser.*, Res. Rep. 39, 29 pp.

BUSS, K. and MILLER, E. R. (1971). "Considerations for Conventional Trout Hatchery Design and Construction in Pennsylvania," *Progr. Fish-Cult.*, 33(2), pp 86–94.

BUTLER, G. and ISON, H. C. K. (1966). "Corrosion and its Prevention in Waters," London, Leonard Hill, 281 pp.

CACES-BORJA, P. and RASALAN, S. B. (1969). "A Review of the Culture of Sugpo, *Penaeus monodon* Fabricius, in the Philippines," Proc. World Sci. Conf. on the Biology and Culture of Shrimps and Prawns, Mexico City, 1967, Ed. M. N. Mistakidis, *FAO Fish Rep.*, 57(2), pp 111–123.

CAMBEFORT, H. (1966). "Les ouvrages ancres au sol," *Travaux*, No. 376, May 1966.

CAMPBELL, H. S. (1969). "The Compromise between Mechanical Properties and Corrosion Resistance in Copper and Aluminium Alloys for Marine Applications," *Ocean Engng*, 1, 387–394.

CAMPBELL, K. C. (1969). "Fish Farming Husbandry: Feeding and Induced Spawning," *World Fishing*, 18(3), pp 47–48.

CASTAGNA, M. A. (1970). "Hard Clam Culture Method developed at VIMS," *Virginia Institute of Marine Science, Marine Resources Advisory Series*, No. 4 (June), 4 pp.

CHANLEY, P. and NORMANDIN, R. F. (1967). "Use of Artificial Foods for Larvae of the Hard Clam, *Mercenaria mercenaria* (L)," *Proc. Natn. Shellfish Ass.*, 57, pp 31–37.

CHAPMAN, C. J. (1964). "Importance of Mechanical Stimuli to Fish Behaviour, Especially to Trawls" in *Modern Fishing Gear of the World, Vol. II,* Fishing News (Books) Ltd, London.

CHAUDESAIGUES, J. (1966). "Les ouvrages maritimes et de navigation intericure en beton precontraint," *Travaux*, No. 376, May 1966.

COLE, H. A. (1968). "The Scientific Cultivation of Sea Fish and Shell Fish," *Journ. Roy. Soc. Arts*, Vol. 116 (June), pp 590–603.

COLE, H. A. (1972). "Marine Fish and Shellfish Cultivation," *Proc. Oceanology International '72 Conf.*, Brighton, 19–24th March, 12 pp.

CONNOLLY, R. A. (1963). "Effect of Seven-year Marine Exposure on Organic Materials," *Mater. Res. Stand*, 3(3), 193.

COOK, H. L. (1966). "Identification and Culture of Shrimp Larvae," *Circ. Fish. Wildl. Serv.*, Wash., 246, pp 12–13.

COOK, H. L. (1969). "A Method of Rearing Penaeid Shrimp Larvae for Experimental Studies," Proc. World Sci. Conf. on the Biology and Culture of Shrimps and Prawns, Mexico City, 1967, Ed. M. N. Mistakidis, *FAO Fish. Rep.*, 57(3), pp 709–715.

COOK, H. L. and MURPHY, M. A. (1966). "Rearing Penaeid Shrimps from Eggs to Postlarvae," *Proc. Southeastern Ass. Game Fish Comm.*, 19th Ann. Conf., pp 283–288.

COOK, H. L. and MURPHY, M. A. (1969). "The Culture of Larval Penaeid Shrimp," *Trans. Amer. Fish. Soc.*, 98, p 751.

COOK, H. L. and MURPHY, M. A. (1971). "Early Developmental Stages of the Brown Shrimp, *Penaeus aztecus* Ives, Reared in the Laboratory," *Fish. Bull.*, U.S. Dept. of Commerce, 69(1), pp 223–240.

DANEL, F., CHAPUE, E., and DHAILLE, R. (1960). "Tetrapods and Other Precast Blocks for Breakwaters," *Journ. Waterways and Harbours Div., ASCE*, Vol. 86, No. WW3, Paper 2590, Sept. 1960.

DANEL, F. and GRESLOU, L. (1962). "The Tetrapod," *Proc. 8th Conf. on Coastal Eng.*, Mexico, Nov. 1962.

DANTEC, J. LE (1953). "L'élevage des Anguilles dans les réservoirs à poissons du Bassin d'Arcachon," *La Pêche Maritime La Pêche Fluviale et La Pêche Pisciculture* (15th May).

DANTEC, J. LE (1955). "Quelques observations sur la biologie des muges des réservoirs de Certes à Audenge," *Rev. Trav. Inst. Pêches Marit.*, 19(1).

DAVIS, H. C. (1969). "Shellfish Hatcheries–Present and Future," *Trans. Am. Fish. Soc.*, Vol. 98, pp 743–750.

DELMENDO, M. N. and RABANAL, H. R. (1956). "Cultivation of 'Sugpo' (Jumbo Tiger Shrimp), *Penaeus monodon* Fabricius in the Philippines," *Proc. Indo-Pacif. Fish. Coun.*, 6(2–3), pp 424–431.

DETHLOFF, J. (1964). "Problems of Electro-fishing and their Solution," in *Modern Fishing Gear of the World*, Vol. II", Fishing News (Books) Ltd, London.

DEWITT, J. W. (1969). "The Pond, Lagoon, Bay, Estuary, and Impoundment Culture of Anadromous and Marine Fishes, with emphasis on the Culture of Salmon and Trout, along the Pacific Coast of the United States," *U.S. Dept. of Comm.*, Technical Assistance Project, 36 pp.

DILL, W. A. (1967). *Proc. World Symp. Warm Water Pond Fish Culture, Fish. Rep.*, 44(1) (i).

DJAJADIREDJA, R. (1970). "Shrimp Culture in Indonesia," *FAO Fish Cult. Bull.*, 2(4), pp 5–6.

DUHOUX, L. (1964). "Fermeture de la Rance–deroulement des travaux et analyse des observations," *La Houille Blanche*, No. 4, 1964.

EDSON, Q. A. and BANYS, R. J. (1971). "Cowlitz Salmon Hatchery," *Jour. Power Div. ASCE*, 97(PO1), Paper No. 7801, pp 61–76.

EDWARDS, E. (1970). "General Programme—Resource Development Section 1970/71," *Irish Sea Fisheries Board Report*, May, 5 pp.

ELDERFIELD, H., THORNTON, L., and WEBB, J. S. (1971)· "Heavy Metals and Oyster Culture in Wales," *Mar. Poll. Bull.*, 2(3), pp 44–47.

ELLIOTT, F. E. (1969). "On the Status of Electric Fishing," *Proc. Oceanology International Conf.*, Brighton, 1969.

ENAMI, S. (1960a). "Studies on the Bubble Net–I," *Kagoshima Univ., Faculty of Fisheries*, Bull. No. 8, 1960.

ENAMI, S. (1960b). "Studies on the Bubble Net–II," *Bull. Jap. Soc. Sci. Fish*, 26(3), 1960.

FAO (1966). "Fishing with Electricity," *Proc. FAO, Symposium*, Belgrade 1966, Pub. by Fishing News (Books) Ltd, 304 pp.

FAO (1969). "Provisional Indicative World Plan for Agricultural Development," *FAO Conf. Document*, C69/4, Vols 1 and 2, 672 pp.

FAO (1970a). *FAO Fish Cult. Bull.* 2(3), pp 1–2.

FAO (1970b). "FAO Technical Conference on Marine Pollution and its Effects on Living Resources and Fishing," *FAO Fish. Tech.*, Paper No. 99, 85 pp.

FAO (1971). "FAO Yearbook of Fishery Statistics," FAO, Rome.

FAVRETTO, L. (1968). "Commercial Aspects of Mussel Culture in the Gulf of Trieste," *Bull. Soc. Adriat. Sci. Nat.*, 56(2), pp 243–261.

FIELDING, J. R. (1966). "New Systems and New Fishes for Culture in the U.S.," *FAO, World Symp. on Warm Water Pond Fish Culture*, FR: VIII/R-2.

FINUCANE, J. H. (1968). "Faunal Production Project," *Bur. Comm. Fish. Biol. Lab.*, Cont. No. 49, pp 11–15.

FINUCANE, J. H. (1969). "Faunal Production Project," *Bur. Comm. Fish. Biol. Lab.*, Cont. No. 55, pp 11–13.

FINUCANE, J. H. (1970a). "Pompano Mariculture in Florida," *Amer. Fish Farm.* (March), pp 5–10, also front and back cover.

FINUCANE, J. H. (1970b). "Pompano Mariculture in Florida," *Proc. Mar. Tech. Soc. Sym.*, "Food–Drugs from the Sea", pp 135–143.

FINUCANE, J. H. (1971a). "Larval Fish Culture," *World Maricult. Soc. Ann. Meet.*, 2, 8 pp.

FINUCANE, J. H. (1971b). "Progress in Pompano Mariculture in the United States," *World Maricult. Soc. Workshop*, 2, pp 69–72.

FORSTER, J. R. M. (1970). "Further Studies on the Culture of the Prawn, *Palaemon serratus* (Pennant) with Emphasis on the Post-Larval Stages," *Fishery Invest. Lond.* (2), 26(6), H.M.S.O., London, 40 pp.

FOSTER, R. F. and DAVIS, J. J. (1956). "The Accumulation of Radioactive Substances in Aquatic Forms," *Proc. Int. Conf. Peaceful Uses of Atomic Energy*, 13, Sess 18c2, pp 364–367.

FREYTAG, G. and KARGER, W. (1969). "The Problem of the Influence of Technical Noises on Fish," *Proc. Oceanology International Conf.*, Brighton, 1969.

FUJINAGA (HUDINAGA), M. (1969). "Kuruma Shrimp (*Penaeus japonicus*) Cultivation in Japan," Proc. World. Sci. Conf. on the Biology and Culture of Shrimps and Prawns, Mexico City, 1967, Ed. M. N. Mistakidis, *FAO Fish. Rep.* 57(3), pp 811–832.

FUJIYA, M. (1965). "Physiological Estimation of the Effects of Pollutants upon Aquatic Organisms," In: Advances in Water Pollution Research. Proceedings of the 2nd International Conference held in Tokyo, 1964, Vol. 3, Ed. by E. A. Pearson, Pergamon Press, Oxford.

FUJIYA, M. (1970). "Oyster Farming in Japan," *Hel. wiss. Meeresunters*, 20 (1–4), pp 464–479.

GALTSOFF, P. S. (1964). "The American Oyster, *Crassostrea virginica*," *U.S. Wildl. Ser. Fish. Bull.*, Vol. 64, 480 p.

GAMLEN (1971). "New Boom Keeps Spillage Oil in Confined Areas," *Manufacturing Management* (Jan.), 2 pp.

GARSTANG, W. (1905). *Mar. Biol. Ass. Internat. Rept. I*, Cd-2670.

GAUCHER, T. A. (1970). "Thermal Enrichment and Marine Aquiculture," in *"Marine Aquiculture"*, Ed. W. J. McNeil, Oregon State University Press, pp 141–152.

GOLIKOV, A. N. and SCARLATO, O. A. (1970). "Abundance, Dynamics, and Production Properties of Populations of Edible Bivalves, *Mizuhopecten yessoensis* and *Spisula sachalinensis* related to the Problem of Organisation of Controllable Submarine Farms at the Western Shores of the Sea of Japan," *Hel. wiss. Meeresunters*, 20(1–4), pp 498–513.

GROSS, F., RAYMONT, E. G., MARSHALL, S. M. and ORR, A. P. (1949). "A Fish-Farming Experiment in a Sea-Loch," *Nature*, London, 153 (Apr.), p. 483.

GROSS, F., RAYMONT, E. G., MARSHALL, S. M., and ORR, A. P. (1949). "An Experiment in Marine Fish Cultivation," *Proc. Roy. Soc. Edinb. B.*, 63, 1947–49, pp 1–95.

GROSS, F., RAYMONT, E. G., GAULD, D. T., MARSHALL, S. M., and ORR, A. P. (1952). "A Fish Cultivation Experiment in an Arm of a Sea-Loch," *Proc. Roy. Soc. Edinb. B.*, 64, 1949–52, pp 1–135.

HALL, D. N. F. (1962). "Observations on the Taxonomy and Biology of some Indo-West-Pacific Penaeidae (Crustacea, Decapoda)," *Fishery Publs. colon. off*, (17), 229 pp.

HALL, J. V. (1968). "Stability Tests of Interlocking Block Revetment," *Journ. Waterways and Habour Div., ASCE*, Vol. 94, No. WW3, Paper No. 6079, Aug. 1968.

HALSBAND, E. (1959). "The Effect of Pulsating Electrical Current on Fish," in *"Modern Fishing Gear of the World*, Vol. 1", Fishing News (Books) Ltd, London, pp 575–580.

HARADA, T. (1965). "Studies on Propogation of Yellowtail (*Seriola quinqueradiata* T. et S.), with Special Reference to Relationship between Feeding and Growth of Fish Reared in Floating Net Crawl," *Mems. Fac. Agric. Kinki Univ.*, 3, pp 1–291.

HARADA, T. (1970). "The Present Status of Marine Fish Cultivation Research in Japan," *Hel. wiss. Meeresunters.*, 20, pp 594–601.

HARRIS, V. E. (1953). "Some Practical Aspects of Electrical Fishing," *Atlantic Fisherman* (Feb.), 1953.

HAVINGA, B. H. (1956). "Mussel Culture in the Dutch Waddensea," *Rapports et Proces-Verbaux, Conseil Permanent International pour L'Exploration de la Mer*, Vol. 140, (3), pp 49–52.

HAVINGA, B. H. (1964). "Mussel Culture," *Sea Frontiers*, Vol. 10, (3), pp 155–161.

HAYWOOD, K. H. and CURR, C. T. W. (1970). "An Operational Research Study into the Economics of Seed Mollusc Production," *Proc. Shellfish Conf. 6–7, April, 1970*, Shellfish Association of G.B., 10 pp.

HELA, I. and LAEVASTU, T. (1970). "Fisheries Oceanography," Fishing News (Books) Ltd, London, 238 pp.

HEMPEL, G. (1970). "Fish-farming, including Farming of Other Organisms of Economic Importance," *Hel. wiss. Meeresunters*, 21, pp 445–465.

HESTER, F. J. (1966). "Man in the Sea and Fisheries of the Future," Exploiting the Ocean, *Trans. 2nd Ann. M.T.S., Conf.* (June), pp 524–529.

HICKLING, C. F. (1962). "Fish Culture," Faber and Faber, London, 295 pp.

HICKLING, C. F. (1968). "The Farming of Fish," Biology in Action Series, Pergamon Press, 85 pp.

HICKLING, C. F. (1970). "Estuarine Fish Farming," *Adv. Mar. Biol.*, Vol. 8, pp 119–213.

HISAOKA, M., NOGAMI, K., TAKEUCHI, O., SUZUKI, M., and SUGIMOTO, H. (1966). "Studies on Sea Water Exchange in Fish Farm II Exchange of Sea Water in Floating Net," *Bull. Naikai Reg. Fish. Res. Lab.*, Cont. No. 115, pp 21–43.

H.M.S.O. (1969). "Fish Toxicity Tests," *Min. Ag. Fish. Food*, H.M.S.O., London.

H.M.S.O. (1970). "Third Report of the Research Committee on Toxic Chemicals," *Aq. Res. Council*, H.M.S.O., London.

HORBUND, H. M. and FREIBERGER, A. (1970). "Slime Films and their Role in Marine Fouling: A Review," *Ocean Engng*, Vol. 1, pp 631–634.

HOUSTON, R. B. (1949). "German Commercial Electrical Fishing Device," *U.S. Dept. Int. Fish and Wildlife Ser., Fish. Lefl.* 348.

HUDINAGA, M. (1942). "Reproduction, Development and Rearing of *Penaeus japonicus* Bate," *Japan J. Zool.*, Vol. 10, pp 305–393.

HUDINAGA, M. and KITTAKA, J. (1967). "The Large Scale Production of the Young Kuruma Prawn, *Penaeus japonicus* Bate," *Inf. Bull. Planktol, Japan*, December issue, Commemoration No. of Dr Y. Matsue, pp 35–46.

HUDINAGA, M. and MIYAMURA, M. (1962). "Breeding of the 'Kuruma' Prawn (*Penaeus japonicus* Bate)," *J. Oceanogr. Soc. Japan*, 20th Ann. Vol., pp 694–706.

HUNTER, C. J. and FARR, W. E. (1970). "Large Floating Structure for Holding Adult Pacific Salmon (*Oncorhynchus* spp)," *J. Fish. Res. Bd. Canada*, 27, pp 947–950.

HYDRAULICS RESEARCH (1964). "Rip-Rap Protection for Slopes subject to Wave Attack," *Hydraulics Research*, Wallingford, 1964.

I.C.I. FIBRES LTD (1968). "Physical properties of I.C.I. Synthetic-fibre Fishing Gear," *Tech. Rep., I.C.I. (Harrogate)* (tD3/4), 5 pp.

IDYLL, C. P. (1965). "Shrimp Nursery: Science explores new ways to Farm the Sea," *Nat. Geogr. Mag.*, 127(5), pp 636–659.

IDYLL, C. P., TABB, D. C., and YANG, W. T. (1969). "Experimental Shrimp Culture in Southeast Florida," *Proc. Gulf. Caribb. Fish. Inst.*, 1968, 21 p 136 (Abstract).

INCE, S. (1962). "Air Bubbles for Protecting Wharf Structures in the Arctic," *Proc. 8th Conf. Coastal Engr*, Mexico (Nov.).

INOUE, H. (1965). "Sea Water Exchange Flow and Stocking Density Rates in Setting-up Shallow Water Fish Farms," *Marine Products Propogation*, Special Issue, No. 4, Japan.

INOUE, H., TANAKA, Y., and SAITO, M. (1966). "On the Water-exchange in the Shallow Marine Fish Farm: Hitsuishi Fish Farm of Hamachi," *Bull. Jap. Soc. Sci. Fish.*, 32(5), pp 384–392.

INOUE, H., TANAKA, Y., and FUKUDA, K. (1970). "On Water Exchange in a Shallow Marine Fish Farm–II: Hamachi Fishfarm at Tanoura," *Bull. Jap. Soc. Sci. Fish.*, 36(8), pp 776–782.

IVERSEN, E. S. (1968). "Farming the Edge of the Sea," Fishing News (Books) Ltd, London, 304 pp.

IVERSEN, E. S. and BERRY, F. H. (1969). "Fish Mariculture: Progress and Potential," *Proc. Gulf. Caribb. Fish. Inst.*, 21, pp 163–176.

IZBASH, S. V. and KHALDRE, KH. YU (1970). "Hydraulics of River Channel Closure," Butterworths, London, 174 pp.

JACHOWSKI, R. A. (1964). "Interlocking Precast Concrete Block Seawall," *9th Conf. on Coastal Eng.*, Lisbon, 1964.

JOYNER, T. (1970). "Pen Rearing of Salmon," *FAO Fish Cult. Bull.*, 2(3), p 7.

KAWAKAMI, T. (1959). "Development of Mechanical Studies of Fishing Gear," *H. Kristjonsson ed. Modern Fishing Gear of the World*, London, Fishing News (Books) Ltd, 175–184.

KAWAKAMI, T. (1964). "The Theory of Designing and Testing Fishing Nets in Model," *H. Kristjonsson ed., Modern Fishing Gear of the World*, 2, London, Fishing News (Books) Ltd, 471–489.

KENNEY, N. T. (1968). "Sharks, Wolves of the Sea," *Nat. Geogr. Mag.*, 133(2) (Feb.), 1968.

KENSLER, C. B. (1970). "The Potential of Lobster Culture," *Amer. Fish Farmer*, Oct. (Reprint), 7 pp.

KERR, N. M. (1970). "Harvesting of Marine Biological Resources by Dredging," *Proc. Inst. Mar. Eng. Symp. Ocean Eng. Section*, Glasgow, 3rd June, pp 14–23.

KESTEVAN, G. L. (1941). "The Biology and Cultivation of Oysters in Australia," *CSIRO, Division of Fisheries*, Report No. 5, pp 1–32.

KEY, D. (1970). "Further Results from Experiments on the Culture of Small Oysters," *MAFF, Shellfish Inf. Leafl.*, No. 17, 12 pp.

KINNE, O. (1970). "Cultivation of Marine Organisms and its Importance for Marine Biology," *Hel. wiss. Meeresunters*, 20, pp 1–5 and pp 707–710.

KJELSTRUP, S. (1963). "Tetrapods North of the Arctic Circle," *Dock and Harbour Authority*, Vol. 44, No. 516, Oct. 1963.

KNOWLES, J. T. C. (1968). "Mollusc Farming," *Proc. Meeting Society for Underwater Technology, London*, 5th Nov. 1968, pp 7–11.

KNOWLES, J. T. C. (1971). "A Method for the Large Scale Culture of Algae," *Underwater Journal*, 3(4), pp 163–165.

KNOWLES, J. T. C. (1972). "Mollusc Cultivation in Britain—The Potentialities and the Problems," *Proc. Conf. Oceanology International '72*, Brighton, pp 84–86.

KOBAYASHI, K., IGARASHI, S., ABIKA, Y., and HAGASHI, K. (1959). "Studies on the Air Screen in Water, 1—Preliminary Observations of Behaviour of a Fish School in Relation to an Air Screen," *Japanese Fisheries Leaflet*, 1959.

KOW, T. A. (1969). "Prawn Culture in Singapore," Proc. World Sci. Conf. on the Biology and Culture of Shrimps and Prawns, Mexico City, 1967, Ed. M. N. Mistakidis, *FAO, Fish. Rep.*, 57(2), pp 85–93.

KUROKI, T. (1959). "Electrical Fishing in Japan," in "*Modern Fishing Gear of the World*, Vol. I", Fishing News (Books) Ltd, London.

LABRID, C. (1969). "L'ostréiculture e le bassin d'Arcachon," Feret et Fils, Bordeaux, France, 215 p.

LARSEN, J. (1960). "Pneumatic Barrier against Salt Water Intrusion," *Jour. Waterways and Harbours Div. ASCE*, Paper No. 2600, 86(WW3), pp 49–61.

LAQUE, F. L. (1969). "Deterioration of Metals in an Ocean Environment," *Ocean Engng*, 1, 299–312.

LETHLEAN, N. G. (1953). "An Investigation into Design and Performance of Electric Fish Screens and Electric Fish Counter," *Trans. Roy. Soc. Edinb.*, 62(2), 1952–54, pp 479–527.

LINT, M. J. W. DE (1964). "The Dam through the Veere Gat," *Asphalt Institute Inf. Series*, No. 131 (15-131), Sept. 1964.

LITTLE, A. H. (1964). "The Effect of Light on Textiles," *J. Soc. Dyers Colour*, 80, 527–534.

LITTLE, A. H. and PARSONS, H. L. (1967). "The Weathering of Cotton, Nylon and Terylene Fabrics in the United Kingdom," *J. Text. Inst.*, 58, 449–462.

LODER, R. T. and ERHO, M. W. (1971). "Wells Hydroelectric Project Fish Facilities," *Jour. Power Div. ASCE*, 97(PO2), Paper No. 7982, pp 301–316.

LOOSANOFF, V. L. (1960). "Recent Advances in the Control of Shellfish Predators and Competitors," *Proc. Gulf. and Carib. Fish. Inst.*, 13th Ann. Ses., pp 113–127.

LOOSANOFF, V. L. and DAVIES, H. C. (1963). "Rearing of Bivalve Larvae," in "*Advances in Marine Biology*", 1, pp 1–136, Academic Press, London.

LOUISIANA WILD LIFE & FISHERIES COMMISSION (1968). "Pond Studies," *Louisiana Wild Life and Fisheries Commission*, 12th Biennial Rept, 1966–67, pp 181–182.

LUCAS, C. E. (1966). "Fish Cultivation," The Intensive Production of Fish for Human Food, *Proc. of Nutrition Soc.*, Vol. 25.

MCKEE, A. (1967). "Farming the Sea," Souvenir Press, London, 314 pp.

MCNEIL, W. J. (1971a). "Acclimatization of Salmon Fry to Sea Water," *FAO Aquacult. Bull.*, 3(4), p 5.

MCNEIL, W. J. (1971b). "Chum Salmon Hatchery Legislation in Oregon," *FAO Aquacult. Bull.*, 3(4), p 11.

M.A.F.F. (1961). "The Purification of Oysters in Installations Using Ultra-violet Light," *Lab. Leafl. Fish. Lab.*, *Burnham-on-Crouch* (Old Series), No. 27, 8 pp.

M.A.F.F. (1966). "Lobster Storage and Shellfish Purification," *Lab. Leafl. Fish. Lab.*, *Burnham-on-Crouch* (New Series), No. 13, 15 pp.

M.A.F.F. (1969). "The production of clean shellfish", *Lab. Leafl. Fish. Lab.*, *Burnham-on-Crouch* (New series), No. 20, 16 pp.

MALONE, T. C. (1969). "Primary Productivity in a Hawaiian Fishpond and Its Relationship to Selected Environmental Factors," *Pacif. Sci.*, 23(1), pp 26–34.

MARSHALL, H. L. (1970). "Development and Evaluation of New Cultch Materials and Techniques for Three-dimensional Oyster Culture, *M.T.S. Journal*, 4(1), pp 7–21.

MARVIN, K. T. (1964). "Construction of Fibreglass Water Tanks," *Prog. Fish. Cult.*, 26(2), pp 91–92.

MASON, J. (1967). "Money from Old Rope," *Scot. Fish. Bull.*, No. 27 (June), 2 pp.

MASON, J. (1969). "Mussel Raft Trials Succeed in Scotland," *World Fishing*, Vol. 18(4), pp 22–24.

MASON, J. (1971). "Mussel Cultivation," *Underwater Journal*, Vol. 3(2), pp 52–59.

MATTHIESSEN, G. C. (1969). "Seed Oyster Production in Fisheries Island, New York," *FAO Fish. Cult. Bull.*, 2(1), p 12.

MAWDESLEY-THOMAS, L. E. (1971). "Toxic Chemicals—the Risk to Fish," *New Scientist, Lond.*, 49(734), pp 74–75.

MEANEY, R. A. (1970). "Mussels in Ireland," *Irish Sea Fisheries Board, Resource Development Note*, July, 12 pp.

MEIXNER, R. (1971). "Wachstum und Ertrag von *Mytilus edulis* bei Flobkultur in der Flensburger Förde," *Arch. Fisch. Wiss.*, 22(1), pp 41–50.

MILLS, D. (1971). "Salmon and Trout," Oliver and Boyd, Edinburgh, 351 pp.

MILNE, P. H. (1969a), "Civil Engineering Aspects of Fish Farming," *Proc. Challenger Soc.*, Vol. 4(1), p 20.

MILNE, P. H. (1969b). "Scaffold Tube in Fish Farm Research," *University Equipment*, Dec. 1969, p 11.

MILNE, P. H. (1970a). "Air Bubble Curtains and their Use in Coastal Waters," *Fluid Power International*, Vol. 35, No. 410, May 1970, pp 33–36.

MILNE, P. H. (1970b). "Fish Farm Enclosures," publication in booklet form of eight articles appearing in *World Fishing*, from Vol. 18, No. 12, 1969 to Vol. 19, No. 7, 1970, 20 pp.

MILNE, P. H. (1970c). "Marine Fish Farming in Scotland," *World Fishing*, Vol. 19, No. 9, pp 46–50.

MILNE, P. H. (1970d). "Collapsible Catamaran for Inshore Waters," *Ship and Boat International*, Vol. 23, No. 9, p 40.

MILNE, P. H. (1970e). "Fish Farming: A Guide to the Design and Construction of Net Enclosures," *Marine Research Scotland*, 1970, No. 1, H.M.S.O., Edinburgh, 31 pp.

MILNE, P. H. (1970f). "Marine Fish Farming Enclosures," *Scottish Fisheries Bulletin*, No. 34 (Dec. 1970), pp 12–14.

MILNE, P. H. (1971a). "Fish from the Farm," *Scots Magazine* (New Series), Vol. 95, No. 2 (May), pp 128–136.

MILNE, P. H. (1971b). "Storm Surge Research on Scottish West Coast," *Dock and Harbour Authority*, Vol. 52, No. 610 (August), pp 150–152.

MILNE, P. H. (1971c). "Hydrographic Research in Enclosed Waters," *Hydrospace*, Vol. 4, No. 5 (October), pp 48–52 and Vol. 4, No. 6 (Dec.), pp 46–48.

MILNE, P. H. (1972a), "Sectional Tanks for Fish Farms," *World Fishing*, Vol. 21, No. 1, pp 20, 22.

MILNE, P. H. (1972b). "Compressed Air Automatic Fish Feeders," *Compressed Air Magazine*, Vol. 77, No. 5.

MILNE, P. H. (1972c). "Fish Farm Engineering," *Proc. Symp. Oceanology International '72*, Brighton, pp 95–99.

MILNE, P. H. (1972d). "Hydrography of Scottish West Coast Sea Lochs," *Marine Research Scotland* (in press), H.M.S.O., Edinburgh.

MILNE, P. H. and POWELL, H. T. (1972). "Marine Fouling on Fish Netting Test Panels at Four Sites on the West Coast of Scotland," to be published.

MINAUR, J. (1971). "A Simple Electrically-operated Feeder for use in Salmon Rearing," *Jour. Fish Biol.*, 3(4), pp 413–416.

MITCHELL, N. T. (1971). "Radioactivity in Surface and Coastal Waters of the British Isles, 1970," *Min. Ag. Fish. Food, Fish. Radiobiol. Lab.*, Tech. Rep. FRL8, 35 pp.

MOE, M. A., LEWIS, R. H., and INGLE, R. M. (1968). "Pompano Mariculture: Preliminary Data and Basic Considerations," *Florida Bd. Cons.*, Tech. Ser. No. 55, 65 pp.

MUNDEY, G. R. (1969). "Highlands Lobster Farm in Operation," *World Fishing*, 18(9), pp 38–39.

MURRAY, P. J. (1971). "Mussel Culture," *FAO Aquacult. Bull.*, 3(4), p 8.

NAGAI, S. (1962). "Stable Concrete Blocks on Rubble Mound Breakwaters," *Journ. Waterways and Harbours Div.*, ASCE, Vol. 88, No. WW3, Paper 3230, August 1962.

NASH, C. E. (1968). "Power Stations as Sea Farms," *New Scientist, Lond.*, 40 (14th Nov.), pp 367–369.

NASH, C. E. (1970). "Marine Fish Farming," *Mar. Poll. Bull.*, 1(NS), No. 1 (Jan.), No. 2 (Feb.).

NIKOLIC, M. (1970). "Oyster Culture Experiments in Cuba," *FAO Fish Cult. Bull.*, 2(4), pp 7–8.

NOMURA, N. and MORI, K. (1956). "Resistance of Plane Net against Flow of Water. III. Effect of Kind of Fibres on the Resistance of the Net", *Bull. Jap. Soc Sci. Fish.*, 21, 1110–1113.

NORTENE (1969). "Le Netlon en Conchyliculture," Nortene, 126 Bd. A. Blanqui, Paris 13.

NOVOSTI INFORMATION SERVICE (1967). "Talking to Fish," Bull. No. 4338 (9th Jan.).

NOWAK, W. S. W. (1970). "The Marketing of Shellfish," Fishing News (Books) Ltd, London, 263 pp.

OKABAYASHI, S. (1958). "On the Deformation of Fishing Net Caused by the Current," *Bull. Jap. Soc. Sci. Fish.*, 24(4), pp 263–266.

OLIVIER, H. (1967). "Through and Overflow Rockfill Dams—New Design Techniques," *Proc. Inst. Civ. Engrs*, Paper No. 7012, Vol. 36, March 1967.

PAAPE, A. and WALTHER, A. W. (1962). "Akmon Armour Unit for Covering Layers of Rubble Mound Breakwaters," *Proc. 8th Conf. on Coastal Eng.*, Mexico, Nov. 1962.

PANKHURST, R. C. and HOLDER, D. W. (1952). "Wind-tunnel Technique," London, Sir Isaac Pitman & Sons Ltd, 702 pp.

PARISOT, T. J. (1967). "A Closed Recirculated Seawater System," *Progr. Fish Cult.*, 29(3), pp 133–139.

PAZ-ANDRADE (1968). "Raft Cultivation of Mussels is Big Business in Spain," *World Fishing*, Vol. 17(3), pp 50–52.

PHILLIPS, G. (1970). "Dover Sole Culture in Recirculated Water," *FAO Fish Cult. Bull.*, 2(2), p 8.

PHINNEY, L. A. (1966). "A Brief of the Artificial Production Progress for Salmon in Washington State," *Washington State Department of Fisheries*, Progress Report, 16 pp.

PHINNEY, L. A. and KRAL, K. B. (1965). Supplemental Report On: An Economic Evaluation of Washington State Department of Fisheries Controlled Natural Rearing Program for Coho Salmon (*Oncorhynchus kisutch*)," *Washington State Department of Fisheries*, 152 pp.

PINCHOT, G. B. (1970). "Marine Farming," *Sci. Amer.*, 223(6), pp 14–21.

PINNER, S. H. (1966). "Weathering and Degradation of Plastics: Based on a Symposium at the Borough Polytechnic, London," 1963, ed., London, Columbia Press, 131 pp.

PROWSE, G. A. (1963). "Neglected Aspects of Fish Culture," *Curr. Aff. Bull., Indo-Pacific. Fish. Coun.*, Vol., 36, pp 1–9.

PRUGININ, J. and BEN-ARI, A. (1959). "Instructions for the Construction and Repair of Fish Ponds," *Bamidgeh*, 11(1).

PUGNET, L. and CAPITAINE, E., "La coupure du Rhin pour l'amenagement hydro-elec de Rhinau," *La Houille Blanche*, No. 4, 1964.

PURVES, A. H. (1968). "Trout Culture in Australia," *Riv. it. Piscic. Ittiopat–A. 111, N.2, Adrile–Maggio –Giugno*, pp 28–32.

QUAYLE, D. B. (1969). "Pacific Oyster Culture in British Columbia," *Fish. Res. Bd. Canada*, Bull. 169, 193 pp.

RADIONOV, S. I. (1958). "Wave Dissipation by Compressed Air (Pneumatic Breakwater)," *Proc. Conf.*, "Ocean Transport", Moscow.

RASMUSSEN, W. W. and LAURITZEN, C. W. (1953). "Measuring Seepage from Irrigation Canals," *Agricultural Engineer*, May 1953.

REEVE, M. R. (1969a). "Growth, Metamorphosis and Energy Conversion in the Larvae of the Prawn, *Palaemon serratus*," *J. Mar. Biol. Assoc., U.K.*, 49(1), pp 77–96.

REEVE, M. R. (1969b). "The Laboratory Culture of the Prawn, *Palaemon serratus*," *Fishery Invest. Lond.* (2), 26(1): 38 pp.

REEVE, M. R. (1969c). "The Suitability of the English Prawn, *Palaemon serratus* (Pennant) for Cultivation– A Preliminary Assessment," *FAO Fish. Rep.*, 57(3): pp 1067–1073.

REYNOLDS, C. E. (1961). "Reinforced Concrete Designer's Handbook," London, Concrete Publ. Ltd, 362 pp.

RICHARDSON, I. D. (1969). "Mollusc Cultivation," *Proc. Conf. Oceanology International '69*, Brighton, 1969, 8 pp.

RICHARDSON, I. D. (1970). "The Way Ahead for Marine Farms," *Fishing News International*, 9(11), pp 18–20.

RICHARDSON, I. D. (1971). "Development of Techniques for the Cultivation of Marine Fish and Shellfish by the White Fish Authority, United Kingdom," *Conf. Oceanexpo '71*, Bordeaux, Mar. Theme II, Vol. 1, Paper G2-10, 5 pp.

RILEY, J. D. and THACKER, G. T. (1963). "Marine Fish Culture in Britain. 3. Plaice (*Pleuronectes platessa* (L.)) Rearing in Closed Circulation at Lowestoft, 1961," *J. Cons. int. Explor. Mer.*, 28, pp 80–90.

ROGERS, T. H. (1960). "The Marine Corrosion Handbook," New York, London, 297 pp.

ROGERS, T. H. (1968). "Marine Corrosion," London, Newnes, 307 pp.

ROLLEFSEN, G. (1939). "Artificial Rearing of Fry of Sea Water Fish. Preliminary Communication," *Rapp. Cons. Explor. Mer.*, 109, p 133.

ROLLEFSEN, G. (1940). "Utklekking og oppdretting av saltvannsfisk," *Naturen*, 6–7, pp 197–217.

ROSATO, D. V. and SCHWARTZ, R. T., ed. (1968). "Environmental Effects on Polymeric Materials," 2 Vols., London, Interscience Publ., 2216 pp.

RUGGLES, C. P. "Oyster Culture in Canada," *FAO Fish Cult. Bull.*, 1(4), p 6.

RYTHER, J. H. and BARDACH, J. E. (1968). "The Status and Potential of Aquaculture, particularly Invertebrate and Algae Culture," Volumes 1 and 2, *U.S. Dept. of Comm., National Technical Information Service*, PB 177 767, 261 pp and PB 177 768, 225 pp.

SAVILLE, T., jr. (1954). "The Effect of Fetch Width on Wave Generation," *Tech. Memo. U.S. Army Corps Engner, Beach Erosion Bd.* (70), 9 pp.

SCHUSTER, W. H. (1949). "Fish Culture in Brackishwater Ponds of Java," *Indo-Pacific Fisheries Council*, Special Publications, No. 1, India.

SCOTT, M. M. (1968). "Salt Water Fish Farming," *Unilever Quarterly, Lond.*, Vol. 52, No. 296, pp 167–173.

SEDGWICK, S. D. (1966). "Rainbow Trout Farming in Denmark," *Scot. Agric.*, (Autumn), pp 186–190.

SEDGWICK, S. D. (1970). "Rainbow Trout Farming in Scotland–Farming Trout in Salt Water," *Scot. Agric.* (Autumn), pp 180–185.

SERENE, P. (1969). "Aquaculture in France," *FAO Fish Cult. Bull.*, 2(1), p 10.

SHEHADEH, Z. H. (1970). "Plastic Pastures for Grey Mullets," *FAO Fish Cult. Bull.*, 2(4), p 3.

SHEHADEH, Z. H., KUO, J., and MADDEN, W. (1971). "Propogation of the Grey Mullet," *FAO Aquacult. Bull.*, 3(3), p 2.

SHELBOURNE, J. E. (1953). "The Feeding Habits of Plaice Post-larvae in the Southern Bight," *J. mar. biol. Ass.*, U.K., 32, pp 149–159.

SHELBOURNE, J. E. (1955). "Significance of the Subdermal Space in Pelagic Fish Embryos and Larvae," *Nature, Lond.*, 176, pp 743–4.

SHELBOURNE, J. E. (1956a). "The Abnormal Development of Plaice Embryos and Larvae in Marine Aquaria," *J. mar. biol. Ass.*, U.K., 35, pp 177–192.

SHELBOURNE, J. E. (1956b). "The Effect of Water Conservation on the Structure of Marine Fish Embryos and Larvae," *J. mar. biol. Ass. U.K.*, 35, pp 275–286.

SHELBOURNE, J. E. (1957). "Site of Chloride Regulation in Marine Fish Larvae," *Nature, Lond.*, 180, 920–922.

SHELBOURNE, J. E. (1963). "Marine Fish Culture in Britain." 2. "A Plaice Rearing Experiment at Port Erin, Isle of Man, in Open Sea Water Circulation." 4. "High Survivals of Metamorphosed Plaice during Salinity Experiments in Open Circulation at Port Erin, Isle of Man 1961," *J. Cons. Int. Explor. Mer.*, 28, pp 70–79, pp 91–100.

SHELBOURNE, J. E. (1964). "The Artificial Propogation of Marine Fish," in *Adv. Mar. Biol.*, Vol. 2, pp 1–83.

SHELBOURNE, J. E. (1965). "Rearing Marine Fish for Commercial Purposes," *Rep. Calif. coop. oceanic. Fish. Invest.*, 10, pp 53–63.

SHELBOURNE, J. E., RILEY, J. D., and THACKER, G. T. (1963). "Marine Fish Culture in Britain. 1. Plaice Rearing in Closed Circulation at Lowestoft, 1957–1960," *J. Cons. int. Explor. Mer.*, 28, pp 50–69.

SHELLARD, H. C. (1965). Extreme Wind Speeds over the United Kingdom for Periods Ending 1963, *Climatol, Memo. Met. Off. Lond.* (50), 17 pp.

SHLESSER, R. (1971). "Lobster Culture," *FAO Aquacult. Bull.*, 3(4), p 8.

SIMPSON, A. C. (1959a). "The Spawning of the Plaice in the North Sea," *MAFF Fish. Invest. Ser. II*, 22(7), 111 pp.

SIMPSON, A. C. (1959b). "The Spawning of the Plaice (*Pleuronectes platessa*) in the Irish Sea," *MAFF Fish. Invest. Ser. II*, 22(8), 30 pp.

SIMPSON, D. (1966). "Electric Fish Screens in Scotland," in FAO Symposium on "Fishing with Electricity", Fishing News (Books) Ltd, London.

SINDERMANN, C. J. (1966). "Diseases of Marine Fishes," *Adv. Mar. Biol.*, 2, pp 1–89.

SINDERMANN, C. J. (1970a). "Disease and Parasite Problems in Marine Aquiculture," in "*Marine Aquiculture*", Ed. W. J. McNeil, Oregon State Univ. Press, pp 103–134.

SINDERMANN, C. J. (1970b). "Principal Diseases of Marine Fish and Shellfish," Academic Press, London, 369 pp.

SINGH, K. Y. (1968). "Stabit–A New Armour Block," *Proc. 11th Conf. on Coastal Eng., London*, Sept. 1968.

SINTEF (1965). "Air Bubble Curtain Experiments Prove Effective," *Comm. Fish. Rev.* (Sept.), 1965.

SLICHTER, F. B. (1967). "Influences on Selection of the Type of Dam," *Journ. Soil Mech. and Founds. Div., ASCE*, Vol. 93, No. SM3, Proc. Paper 5224, May 1967

SMITH, K. A. (1964). "The Use of Air Bubble Curtains as an Aid to Fishing," in "*Modern Fishing Gear of the World, Vol. II*", Fishing News (Books) Ltd, London.

SORENSEN, J. H. (1970). "Aquaculture in New Zealand," *FAO Fish Cult. Bull.*, 2(4), pp 10–11.

SPOTTE, S. H. (1970). "Fish and Invertebrate Culture–Water Management in Closed Systems," Wiley-Interscience, 145 pp.

STAFF, C. E. (1967). "Seepage Prevention with Impermeable Membranes," *Civil Engineering ASCE*, Vol. 37, No. 2, Feb. 1967.

STEELE, J. H. (1968). "Marine Food Chains," (Ed.), Sym. at Aarhus 1968, Oliver & Boyd, London.

SUGIMOTO, H., HISAOKA, M., NOGAMI, K., TAKEUCHI, O., and SUZUKI, M. (1966). "Studies on Sea Water Exchange in Fish Farm–I: Exchange of Sea Water in Fish Farm Surrounded by Net," *Bull. Naikai Reg. Fish. Res. Lab.*, Cont. No. 113, pp 1–20.

SWIFT, D. R. (1968). "Marine Fish Farming," *Proc. Meet. Society for Underwater Technology*, 5th Nov. London.

SWIFT, D. R. (1969). "Fish Farming," *Proc. Conf. Oceanology International '69*, Brighton, 5 pp.

SYMPOSIUM (1964). "Grouts and Drilling Muds in Engineering Practice," Butterworths, London, 1964.

TABB, D. C., YANG, W. T., IDYLL, C. P., and IVERSEN, E. S. (1969). "Research in Marine Aquaculture at the Institute of Marine Sciences, University of Miami," *Trans. Fish. Soc.*, No. 4, pp 738–742.

TAMURA, T. and YAMADA, S. (1963). "Study of the Construction Works of Fish Pond from Civil Engineering Viewpoint," *Booklet mar. Resour. Conserv. Soc. Japan* (1), 32 pp.

TERADA, T., SEKINE, I., and NOZAKI, T. (1914). "Study on the Resistance of Fishing Net against Flow of Water," *J. imp. Fish. Inst., Tokyo*, 10(5), pp 1–23.

THAM, A. K. (1955). "The Shrimp Industry of Singapore," *Proc. Indo-Pacif. Fish. Coun.*, 5(2), pp 145–155.

THOMAS, H. J. (1964). "Artificial Hatching and Rearing of Lobsters–A Review," *Scot. Fish. Bull.*, No. 21, pp 6–9.

THORN, R. B. (1960). "The Design of Sea Defence Works," Butterworths, London, 1960.

U.S. ARMY COASTAL ENGNG RES. CENTER (1966). "Shore Protection, Planning and Design," 3rd ed., *Tech. Rep. U.S. Army coastal Engng Res. Cent.* (4), 561 pp.

VIK, K. O. (1963). "Fish Cultivation," *Salmon and Trout Mag.*, No. 169 (Sept.), pp 203–208.

VILLADOLID, D. V. and VILLALUZ, D. K. (1951). "The Cultivation of Sugpo, *Penaeus monodon*, Fabricius in the Philippines," *Philipp. J. Fish.*, 1, pp 1–3.

WALNE, P. R. (1956). "Experimental Rearing of the Larvae of *Ostrea edulis* in the Laboratory," *MAFF. Fish. Invest. Ser. II*, 20(9), 30 pp.

WALNE, P. R. (1965). "Observations on Influence of Food Supply and Temperature on the Feeding and Growth of the Larvae of *Ostrea edulis* L.," *MAFF. Fish. Invest. Ser. II*, 24(1), 45 pp.

WALNE, P. R. (1966). "Experiments in the Large Scale Culture of the Larvae of *Ostrea edulis* L.," *MAFF. Fish. Invest. Ser. II*, 25(4), 53 pp.

WALNE, P. R. (1969). "Studies on the Food Value of Nineteen Genera of Algae to Juvenile Bivalves of the Genera *Ostrea*, *Crassostrea*, *Mercenaria* and *Mytilus*," *MAFF. Fish. Invest. Ser. II*, 26(5).

WALNE, P. R. (1970a). "Present Problems in the Culture of the Larvae of *Ostrea edulis*," *Hel. wiss. Meeresunters*, 20, pp 514–525.

WALNE, P. R. (1970b). "Culture of Exotic Species of Bivalve," *Proc. Shellfish Conf. 6–7 April 1970*, Shellfish Association of G.B., 5 pp.

WALNE, P. R. (1972). "Hatchery Cultivation of Bivalves," *Proc. Conf. Oceanology International '72*, Brighton, pp 82–83.

WALNE, P. R. and SPENCER, B. E. (1971). "The Introduction of the Pacific Oyster (*Crassostrea gigas*) into the United Kingdom," *MAFF. Shellfish Inf. Leaflet*, No. 21, 14 pp.

WEBBER, H. H. (1968). "Mariculture," *Bio. Sci.*, Vol. 18(10), pp 940–945.

WEBBER, H. H. (1970). "The Development of a Maricultural Technology for the Penaeid Shrimps of the Gulf and Caribbean Region," *Hel. wiss. Meeresunters*, 20(1–4), pp 455–463.

WHEELER, R. S. (1966). "Cultivation of Shrimp in Artificial Ponds," *Circ. Fish Wildl. Serv.*, 246, pp 14–15.

WIBORG, K. F. and BØHLE, B. (1968). "Den spanske blåskjellindustri: dyrking og foredling, samt notater om østersdyrking og skjellgraving i Vigoområdet i Nord-vest Spania," *Fiskets Gang*, Vol. 54(6), pp 91–95.

WIEGEL, R. L. (1965). "Oceanographical Engineering," Englewood Cliffs, N.J., Prentice-Hall, Inc., 532 pp.

WIERSMA, A. G. (1960). "Turfing on Sea Walls," *Proc. Instn. Civ. Engnrs*, 15 (March).

WILLIAMSON, G. R. (1971). "Lessons from Japan in Fish Farming," "Aquaculture in Japan–2", *Fishing News International*, 10 (May), pp 30–35 and 10 (June), pp 32–33.

WISE, L. W. (1970). "Reconditioned Recirculated Water Systems for Fish Rearing Installations," *Proc. Northeast Fish and Wildlife Conf., Engineering Section*, Delaware, 20 pp.

YONGE, C. M. (1966a). "Farming the Sea," *Discovery*, 27(7), July, pp 8–12.

YONGE, C. M. (1966b). "Oysters," Collins, New Naturalist, London.

YONGE, C. M. (1970). "Oyster Cultivation," *Underwater Journal*, 2(3), pp 138–144.

ZAITSEV, V. (1970). "USSR Experiments with Oysters and Sea Plants," *World Fishing*, 19(2), p 38.

ZUIDHOF, J. (1966). "Air Bubble Barriers Stop Salt Water Intrusion," *Civil Engineering, ASCE*, 36(7), p 69.

Glossary

Adult: any animal that has attained full growth or is sexually mature.

Air bubble curtain: a continuous stream of bubbles emitted from a sunken hose perforated at intervals through which compressed air is released.

Algae: fresh water and marine chlorophyll-bearing plants ranging in size from a few microns to many metres in length.

Aquaculture (Aquiculture): see sea farming.

Ascidian: a sea-squirt, or tunicate, a degenerate survivor of the ancestors of the vertebrates, shaped like a double-mouthed flask.

Brackish water: any mixture of sea water and fresh water with a salinity of less than 30 parts per thousand.

Brine shrimp: (*Artemia salina*) a small crustacean which can be easily reared as a food for early stages of fishes and shell-fishes.

Byssus: a mass of horny fibres secreted by a gland in some molluscs, which is used for attachment.

Catchment area: the area from which a river or reservoir is fed due to natural rainfall.

Closed season: a period when fishing for particular species is restricted, with the intention of protecting that species during the breeding season.

Collectors: see cultch.

Competition: struggle between organisms of the same or different species for necessities of life.

Crustaceans: a large class of arthropod animals, almost all aquatic. Examples: crabs, lobsters, shrimps, etc.

Cultch: tiles, old oyster shells, or any material used by oyster farmers to collect the young of oysters (spat).

Cultivate: to raise crops with labour and care. On oyster beds, for example, to rake the oysters to remove predators and break up large clusters for relaying.

Culture: the practice of cultivating, as of the soil or water. Raising plants and animals with a view to their improvement.

Disease: a deranged condition of an organism which may be inherited or caused by parasites, dietary deficiencies, or by physical and chemical factors in the environment.

Dredge: a floating machine for raising molluscs, oysters and mussels, from the seabed.

Drills: snails which destroy molluscs by rasping a hole in their shells and eating the fleshy portion.

Ecology: the study of the relations of organisms to their environment.

Fetch length: the open water distance to nearest land in the direction of the prevailing winds.

Finfish: cold-blooded lower aquatic vertebrates possessing fins, and, usually, scales.

Fiord (Fjord): an arm of the sea extending inland, usually long and narrow and bordered by steep cliffs.

Food web (food chain): transfer of food energy through a series of organisms with many stages of eating and being eaten (e.g. plants are eaten by shrimps, shrimps are eaten by fishes, and so on).

Foreshore: the intertidal beach zone between high and low water marks.

Foster-Lucas pond: a hybrid of circular and raceway ponds, having straight sides and circular ends with average dimensions, $5 \times 23 \times 1$ m. These ponds have a centre baffle wall with the water inlet and outlet on either side.

Gravid: pregnant or ripe, ready to spawn.

Hatchery: a place for artificial hatching, especially of fish and shellfish larvae.

H.W.M.O.S.T.: high water mark ordinary spring tides.

Hydrography: the investigation of seas and other bodies of water, including charting, sounding, study of tides, currents, temperatures, salinity, dissolved oxygen content, etc.

Hydrology: the study of water resources in land areas, expecially the prediction of rainfall and stream flows.

Intertidal zone: area on the foreshore lying between low water spring tides and high water spring tides.

Invertebrates: lower animals, without backbones.

Juveniles: young stages of animals, usually between the post-larval stages up to the time they become sexually mature.

Lab-lab: name given by Filipinos to a complex of aquatic plants which includes algae, bacteria, protozoans, and diatoms. This complex may form a dense mat on the bottom of ponds and is eaten by young milkfish and shrimps.

Larvae: an immature stage of an animal which differs greatly in appearance and behaviour from adults.

Levee: a natural or artificial waterside embankment, constructed to contain water such as in a pond.

Liner: a material, such as polythene sheeting, applied to the earth or sand surface of a pond to retain water and prevent seepage.

L.W.M.O.S.T.: low water mark ordinary spring tides.

Mariculture: see sea farming.

Maritime: pertaining to the sea.

Metamorphose: a change of shape, transformation; as when the symmetrical larvae of plaice flatten and settle on the seabed.

MLW: mean low water; sometimes used when there is little variation between spring and neap tide levels.

Minimum size: the size below which it is illegal for anybody to possess or sell fish or shellfish, except by special permit.

Molluscs: animals with soft body coverings with limey shells of 1 to 8 parts or sections. In some species the shell is lacking or reduced.

Moult: periodic shedding of the outer covering, such as the exoskeleton in the Arthropoda (shrimps, crabs, lobsters, and so on).

201

Mysis: the last larval stages in crustaceans before they transform into juveniles.

Nauplius: first larval stage occurring in many crustacean species (e.g. shrimps); characterised by an unsegmented body and three pairs of appendages.

Neap tide: a tide of minimum amplitude, occurring when the sun, moon and earth are at right angles to one another (1st and 3rd quarter of the moon).

Net barrier: a structure erected in the sea from which retention fish nets are hung to form an enclosure for sea farming.

Off-bottom culture: see raft culture.

Pathogenic: causes disease.

Pesticide: a chemical substance used to kill pests.

Phytoplankton: tiny plants which drift with the currents.

Plankton: tiny plants and animals which drift with the currents.

Pneumatic barrier: see air bubble curtain.

Pollution: specific impairment of water quality by sewage, pesticides, and industrial waste; may create a hazard to public health.

Post larvae: past the larval stages: stages which resemble the juvenile but are still lacking some characters.

Polymeric: man-made plastic materials, such as polyethylene.

Predation: the act of an animal eating another (prey) of a different, and usually smaller, species.

Propogation: to multiply plants and animals by any method from parent stock.

Raceway: a long narrow pond, with average dimensions, $2.5 \times 30 \times 1.0$ m, where the water inlet and outlet are at opposite ends.

Raft culture: growing oysters or mussels on shells or other materials suspended from rafts or floats. The term is sometimes used to describe any method of hanging culture.

Raising: to cause or promote the growth of animals.

Rearing: to care for and support up to maturity, as to raise shrimp to adults.

Relaying: collecting oysters, clams, or mussels in one location and planting them in another to obtain better growth or better quality meats.

Ren: a collector consisting of oyster or scallop shells threaded on a galvanised wire.

Resistant: able to withstand adverse environmental conditions or ward off diseases.

Run-off: stream water, due to rainfall, flowing from the catchment area.

Salinity: the saltness of water, expressed as parts per thousand (‰).

Sea farming: to promote or improve growth and hence production of marine and brackish water plants and animals by labour and attention, at least at some stage of the life cycle, on areas leased or owned. Usually intended as a profit-making venture.

Sea water: water usually with salinity of 30 to 35 ‰, as found in the open oceans. The salinity of estuarine waters is usually below this down to 20‰.

Seed: young animals, generally oysters, clams, or mussels, used to stock ponds.

Sessile: stationary or attached animals such as mussels or sponges.

Shellfish: aquatic invertebrates possessing a shell or exoskeleton, usually molluscs or crustaceans.

Shuck: to remove shells from oysters, clams, etc., for market, or in preparation for eating.

Sluice gate: a structure with a gate for regulating the inflow or outflow of water.

Spat: young oysters just past the veliger stage which have settled down and become attached to some hard object.

Spat-fall: the settling or attachment of young oysters which have completed their larval stages.

Spring tides: a tide of maximum amplitude, occurring when the sun, moon and earth are in a straight line (at new or full moon).

Storm surge: a rise in tidal height, above the predicted tide, due to wind blowing onshore.

Sublittoral zone: shallow, inshore areas, usually depths down to 200 metres.

Tambak: brackish water pond (Indonesia).

Tonne: Metric ton, 2,204·6 pounds.

Turbidity: a cloudy condition of water, usually caused by impurities. May result from wave action stirring up bottom sediments.

Upwelling: a process by which nutrient-rich bottom waters of the sea are brought near the surface.

Vertebrates: higher animals with backbones (vertebra).

Zoea: a larval stage of some arthropods such as shrimp and crabs.

Zooplankton: tiny animals which drift with the currents.

Index

(a) Scientists and Experts cited

(b) Geographical

(c) Biological

**Other books published by
Fishing News Books Limited
Farnham, Surrey, England**

Free catalogue available on request

A living from lobsters
Advances in aquaculture
Aquaculture practices in Taiwan
Better angling with simple science
British freshwater fishes
Coastal aquaculture in the Indo-Pacific region
Commercial fishing methods
Control of fish quality
Culture of bivalve molluscs
Eel capture, culture, processing and marketing
Eel culture
European inland water fish: a multilingual
  catalogue
FAO catalogue of fishing gear designs
FAO catalogue of small scale fishing gear
FAO investigates ferro-cement fishing craft
Farming the edge of the sea
Fish catching methods of the world
Fish farming international No 2
Fish inspection and quality control
Fisheries of Australia
Fisheries oceanography
Fishery products
Fishing boats and their equipment
Fishing boats of the world 1
Fishing boats of the world 2
Fishing boats of the world 3
Fishing ports and markets
Fishing with electricity
Fishing with light

Freezing and irradiation of fish
Handbook of trout and salmon diseases
Handy medical guide for seafarers
How to make and set nets
Inshore fishing: its skills, risks, rewards
International regulation of marine fisheries: a
  study of regional fisheries organizations
Marine pollution and sea life
Mechanization of small fishing craft
Mending of fishing nets
Modern deep sea trawling gear
Modern fishing gear of the world 1
Modern fishing gear of the world 2
Modern fishing gear of the world 3
Modern inshore fishing gear
More Scottish fishing craft and their work
Multilingual dictionary of fish and fish products
Navigation primer for fishermen
Netting materials for fishing gear
Pair trawling and pair seining—the technology of
  two boat fishing
Pelagic and semi-pelagic trawling gear
Planning of aquaculture development—an
  introductory guide
Power transmission and automation for ships
  and submersibles
Refrigeration on fishing vessels
Salmon and trout farming in Norway
Salmon fisheries of Scotland
Seafood fishing for amateur and professional
Ship's gear 66
Sonar in fisheries: a forward look
Stability and trim of fishing vessels
Testing the freshness of frozen fish
Textbook of fish culture; breeding and
  cultivation of fish
The edible crab and its fishery in British waters
The fertile sea
The fish resources of the ocean
The fishing cadet's handbook
The lemon sole
The marketing of shellfish
The seine net: its origin, evolution and use
The stern trawler
The stocks of whales
Training fishermen at sea
Trawlermen's handbook
Tuna: distribution and migration
Underwater observation using sonar